DESCRIPTIVE CATALOGUE OF THE PATHOLOGICAL SERIES IN THE HUNTERIAN MUSEUM

DESCRIPTIVE CATALOGUE
OF THE PATHOLOGICAL SERIES IN THE HUNTERIAN MUSEUM OF THE ROYAL COLLEGE OF SURGEONS OF ENGLAND

PART II

A selection of surviving specimens
illustrating Hunter's opinions
on the nature of diseases,
experiments and observations
on cases in surgery

E. & S. LIVINGSTONE
EDINBURGH & LONDON
1972

ISBN 0 443 00794 2

Printed in Great Britain

Foreword

It was my privilege to write a foreword to the first volume, published in 1966. It is now my pleasant duty to perform a similar function on the occasion of publication of Part II, the second volume of this descriptive catalogue.

To the Trustees of the Hunterian Collection it has come as a pleasant surprise to be presented with this volume after what has appeared to be a relatively short interval; it is especially welcome since it completes the description of Hunter's pathological specimens, as well as those in the physiological series.

The present volume contains descriptions of all specimens other than those in volume one; the latter was in Hunter's own words. The collection was described by Clift, who had intimate knowledge of Hunter's work, but the present catalogues provide a complete record of the whole series.

For me it remains to express the satisfaction and gratitude of the Hunterian Trustees for Dr Proger's completion of this unique collection; although no longer in daily attendance at the College he has devoted his time to completion of the task and he has made many journeys from his home of retirement on the borders of Dartmoor. In this work he has had much help from Miss Dobson, combined with the painstaking efforts of those who attend to the welfare of the many specimens. The sincere thanks of the Board of Trustees are extended to them all. Dr Proger will, without doubt, gain much satisfaction from completion of his self-imposed task to add to the pleasure of retirement in a peaceful part of the west country, after many years spent in the turmoil of the metropolis.

1972

VICTOR NEGUS
Chairman of the Board of Trustees

Preface

With three exceptions, all the specimens here described originally belonged to John Hunter but there are few surviving records of his views on the nature of the pathological conditions shown or of the order in which he wished to display the preparations. Several specimens have, however, been traced to the original manuscripts and relevant quotations have been inserted. The others have been arranged in a regional manner with recent descriptions. It will be appreciated that the true nature of many of the conditions shown is still uncertain, since the tissues no longer retain their histological staining properties.

Hunter's particular interest in bone is shown by the very large number of specimens in this series and by his important observations and experiments on the growth of bone (see *Physiological Series*, Part I, Nos. 188-201, pp. 8-12). It is known that at the time of his death he was engaged in a detailed review of this section of the museum and recorded his views on the nature of the pathological processes shown. Unfortunately these records were destroyed by Sir Everard Home and only a few of his earlier notes have survived.

1972

JESSIE DOBSON
Curator

Sources and References

Original MS. Catalogue of Hunter's pathological specimens (William Clift's Transcript).

John Hunter's Case Books (William Clift's Transcript):

Vols. I and II	Cases in Surgery.
Vol. III	Dissections of Morbid Bodies.
Vol. IV	Cases and Observations.
Vol. V	Cases and Dissections.

William Clift's Lecture Notes: from *John Hunter's Practical Lectures on Surgery* delivered by Everard Home. 1794, 5 and 6.

That these Lectures are to be considered as Mr. Hunter's is manifest from the following statement:

"In the year 1792, Mr Hunter found that his course of lectures took up so much of this time that he was unable to correct his other papers: he therefore gave it up to me. As a previous step to this arrangement I had given it for the two preceding summers. Mr Hunter now began to prepare the present work for the press and intended as soon as it was in the hands of the Public to give a Course of Practical Lectures in Surgery, for which he had for many years been collecting materials; these were so far advanced, that another winter, had he lived, would have finished them. The materials of these lectures having come into my hands; that they may not be entirely lost to the public, I mean to avail myself of them, and am preparing my arrangements for that purpose."

Hunter: *On the Blood, Inflammation and Gunshot Wounds*;
Life of the Author by Everard Home, p. xxxvi.

James Parkinson's Lecture Notes: from *John Hunter's Lectures on the Principles of Surgery*, delivered in Castle Street. 1785 and 1786. 2 vols.

Hunterian Reminiscences. London, 1833.

Pathological Catalogue. 3rd Edition. 1909. By Professor S. G. Shattock.

The Works of John Hunter, edited by James Palmer. 1835. 4 vols. With vol. of Plates. 1837.

William Clift's copies of original notes and labels on jars.

Sir Everard Home's publications, including transcripts of original MSS.

Matthew Baillie's *Morbid Anatomy of some of the most Important Parts of the Human Body*. London, 1793.

Contents

I

Bone

1. *Malformations and Deformations*
P. 430-450

SKULL

Hyperostosis Frontalis

P. 430. "A vertical section of the calvaria of an adult. The parietal bone is much thickened near its centre, the os frontis remarkably so. (This skull must have been of unusually small dimensions. There is no trace of the coronal suture.)" *MS. Catalogue.*

The divided surface exhibits a perfectly normal structure, in which both the tables and intervening diploë are represented; the outer and inner surfaces of the bone are perfectly smooth. At its posterior margin the frontal is less thickened, and is continuous with the parietal, which in the neighbourhood of the ankylosed coronal suture is almost as much thickened as the more central part of the frontal bone.

The cranial cavity is diminished in the position of the thickenings; the exterior is elevated, but to a much less degree. Drawing No. 23, fig. 1.

P. 431. Portion of a frontal bone increased in thickness by the growth of healthy diploë. (See P. 436.)

P. 432. The upper part of a skull (originally marked "Mr. Ferguson"), increased in weight and thickness. The chief increase is in the frontal bone. The internal table of the skull is displaced inwards, so that the anterior region of the cranial cavity is diminished to an extent which is not indicated by the form of the unaltered external table. The internal surface of the skull is deeply marked by the vascular grooves and apertures; and on the interior of the frontal bone are several irregular circumscribed depressions, such as might be formed by large Pacchionian glands. There are fewer and smaller depressions of the same kind on the inner surface of the parietal bone. The distinction between the tables and diploë is still recognisable. The sutures are not obliterated.

This specimen does not accord with the description of the case of Mr Ferguson recorded in MS. vol. v (*Cases and Dissections*), p. 34.

P. 433. Part of a calvaria showing thickening of the diploë, chiefly in the parietal bones. The thickening is mainly at the expense of the cranial cavity.

See *Parkinson's Lecture Notes*, vol. 2, p. 43.

Drawing No. 22, of which the description is as follows:

"The solid substance of a bone is as subject to swell from disease as the other parts, and the causes seem to be the same in both. They swell from inflammation and they have their indolent swellings, which, when not circumscribed are, I suspect, chiefly scrofulous: at least I have often seen swelling of the bones in persons otherwise scrofulous."

P. 434. A calvaria from an adult, which is increased in thickness and sclerosed. The chief thickening involves the frontal and parietal regions. The coronal and sagittal sutures are obliterated internally. The distinction between diploë and tables is clearly marked.

It also has a singular os triquetrum in the sagittal suture.

P. 435. The right half of a skull, the upper and lateral walls of which are generally and almost uniformly thickened. The greatest thickness is at the middle of the right half of the frontal bone, the section of a portion of which, cut out with a trepan, measures $\frac{1}{2}$ inch. The increase is chiefly owing to the augmentation of the diploë, and is accompanied by a decrease in the capacity of the skull. The texture of the diploë is hard and heavy, and the cancellous spaces are small. The tables of the skull are thick, somewhat rough and porous; and their lines of boundary from the diploë are obscure. The interior is deeply marked with vascular grooves and apertures.

P. 436. A small circular piece of a skull, removed by a trephine, showing thickening and sclerosis of the diploë. (See P. 431, which was originally mounted with this specimen but was not, apparently, from the same individual.)

Hydrocephalus Deformity

P. 437. A calvaria, abnormally wide and rounded in form. The external surface of the parietal bones and of the frontal bone in the situation of the eminences is slightly raised by a thickening of the bone. The frontal eminences are, in consequence, abnormally prominent and rounded. The surface of the bones is rough and abnormally porous. At the anterior end of the sagittal suture is an oval Wormian bone. Internally the calvaria shows considerable thickening of the frontal and anterior part of the parietal bones. The changes are probably due to hydrocephalus and rickets in early life.

Spine

Defect of Atlas

P. 438. The posterior part of the base of a skull, with the atlas. The arch of the atlas is defective in the middle line and its left half is ankylosed to the margin of the foramen magnum, forming a complete bony canal for the vertebral artery. The two atlanto-occipital joints are ankylosed, and the anterior arch of the atlas is united by bone in its whole extent to the basilar process. Viewed from above, the foramen magnum is somewhat narrowed in its posterior half, by projection of the lateral margin.

Lateral Curvature

P. 439. Three dorsal vertebrae, the articular processes and arches of which are ankylosed on the right side. There is also a lateral curvature of this part of the spine and the bodies of the vertebrae are atrophied and reduced in thickness in the concavity of the curve.

P. 440. The last cervical and eleven of the dorsal vertebrae of a spine with extreme lateral curvature. The chief distortion is in the six lower dorsal vertebrae, which form a curve, the convexity of which is directed backwards and to the right side, and which are rotated markedly outwards to the right, their spinous processes being turned, but in a less degree, to the same side. The left half of the bodies of the vertebrae are diminished in depth so that in the concavity of the curve the distance between the lower borders of the sixth and eleventh dorsal vertebrae is hardly more than an inch. The articular processes, the arches and the left sides of the bodies of the lower dorsal vertebrae are united by bone. The five upper dorsal and the seventh cervical vertebrae are less diseased: they present a curvature, of which the concavity is turned towards the right, but their bodies are scarcely at all rotated. The spinous and articular processes and the left side of the bodies of the fourth and fifth dorsal vertebrae are united by bone. A part of the seventh rib on the right side is preserved; its head and neck are united by bone to the articular processes and bodies of the two adjacent vertebrae; and its shaft, as far as the angle, lies parallel with the side of the spine. The articular surfaces for the ribs of the left side are enlarged by peripheral bony deposits, especially those in the concavity of the interior curvature.

Everard Home: *Lecture Notes*, 1794-5, vol. i, pp. 336-338.

Kyphosis

P. 441. Part of a vertebral column. There is a marked kyphotic curve in the upper dorsal region, with its centre at the fifth dorsal vertebra. There is also a slight scoliosis in the same region, the bodies of the third to the seventh vertebrae being rotated towards the right side. The normal slight curve on the left of the vertebral bodies is exaggerated in the specimen as if by some undue pressure, and the bone more compact than in other parts of the column. The fifth-sixth vertebral bodies are ankylosed by a scanty deposit of bone at their contiguous edges, and throughout the column the intervertebral discs appear to have been atrophied. All the bones are lighter and more porous than normal, probably from old age. Drawing No. 64.

Kyphoscoliosis

P. 442. The skeleton of the trunk of an adult female. The lower dorsal portion of the spine is much curved, with its convexity directed to the right, and with slight rotation of the bodies of the vertebrae, their left articular processes being turned forwards. A curve exists in the opposite direction in the lumbar portion of the spine. The bodies of the vertebrae and the intervertebral spaces are reduced in depth, and their margins are very prominent on the side directed to the concavity of each curve. The spinous processes of the vertebrae engaged in the upper curve (that to the right) are directed rather to the left, those of the vertebrae in the lower or leftward curve to the right. The ribs are all approximated; the angles of those of the right side, towards which the dorsal part of the spine is curved, are less obtuse than is natural; and this side

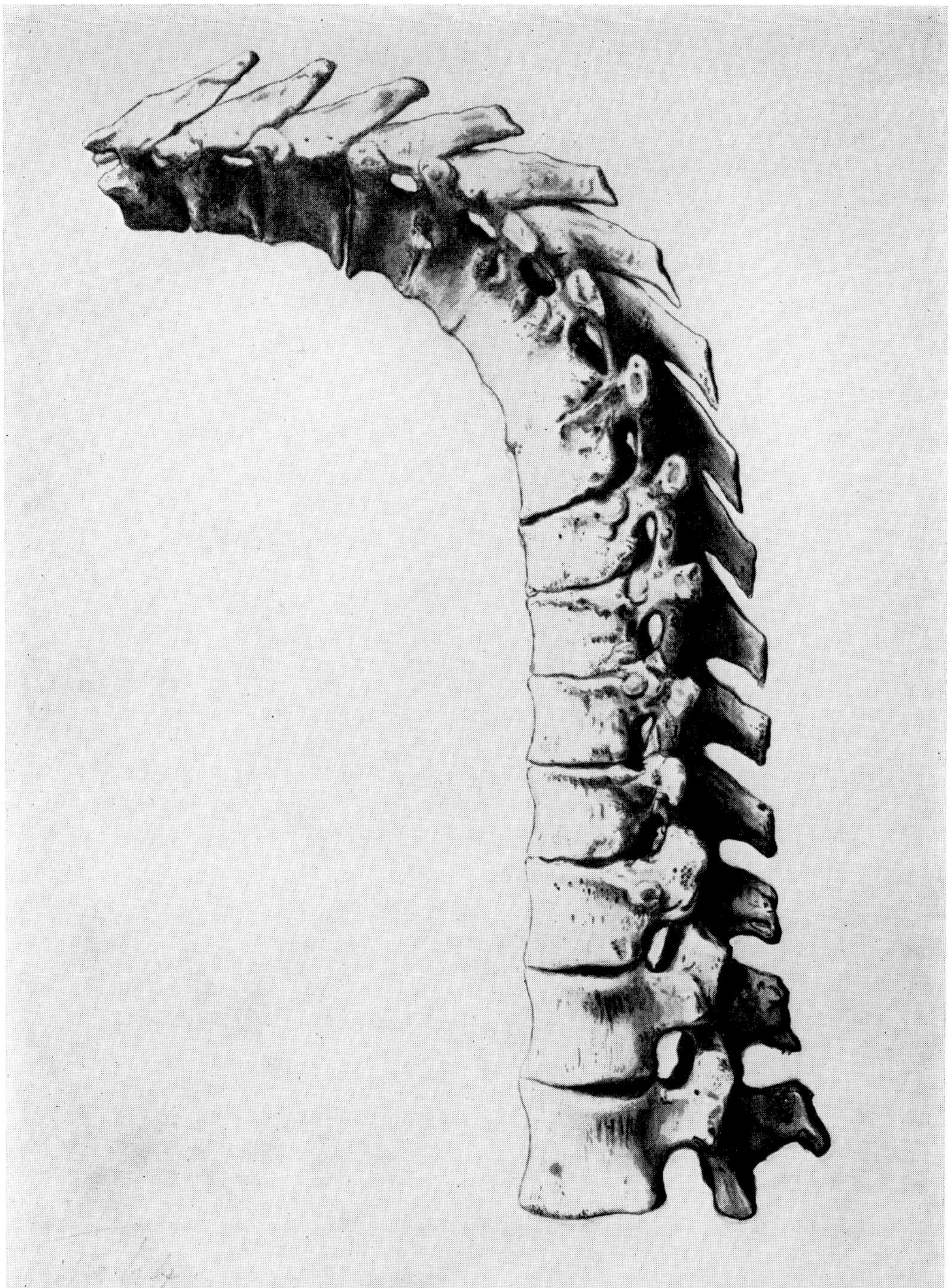

FIG. 1
"Anchylosis of the bodies of five Dorsal Vertebrae, human." Drawing No. 66, by William Bell. Specimen 441.

Face p. 4

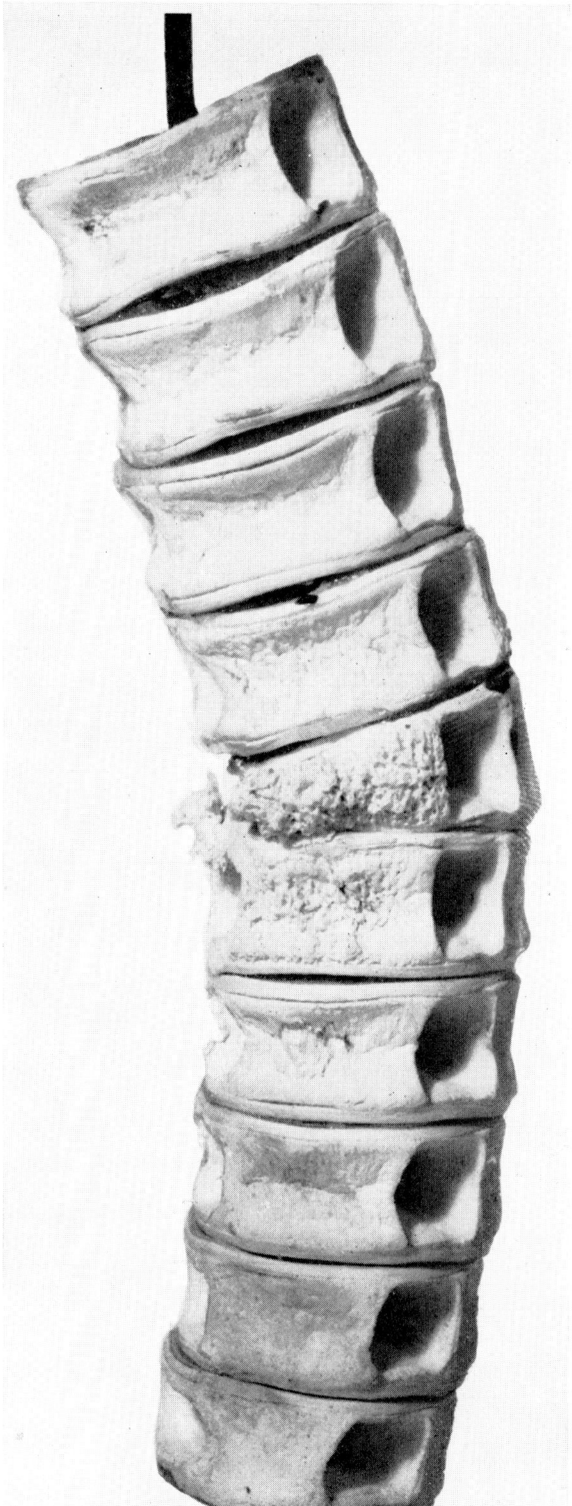

Fig. 2
Spine of a fish deformed by ulceration. Specimen 443.

of the chest is much contracted. On the left side the angles of the ribs are unusually obtuse; and the capacity on this side of the chest is enlarged, at least in its lateral dimensions, nearly enough to compensate for the contraction of the other side. Looked at from behind, the right ribs project prominently backwards at their angles, while those on the left side are flattened and placed further forwards than natural. It will also be seen that the line corresponding to the tips of the spinous processes has the curves much less pronounced than those of the axis of the vertebral bodies. The left transverse process of the fifth lumbar vertebra is large and articulates with the lateral mass of the sacrum. The pelvis is otherwise normal, but is tilted, so that the iliac crest is on a level with the ribs on the right side.

Everard Home: *Lecture Notes*, 1794-5, vol. i, pp. 336, 338. *Duncan's Medical Commentaries*, 8th ed., 1779, vol. vi, pp. 71, 318.

Deformed by Ulceration

P. 443. Part of the spine of a cartilaginous fish, probably a large Ray. One half of the body of a vertebra in the middle of the series has been destroyed, resulting in marked angular curvature. New bone has been formed in small quantity on the bodies of the diseased and the adjacent vertebrae.

Lumbar Assimilation

P. 444. The pelvis and last lumbar vertebra of a Wolf. On the left side the lumbar vertebra is assimilated in form to the first sacral vertebra, exhibiting a well-formed ala, which unites with the sacrum below articulating with the iliac bone and completing the foramen for the passage of the nerve on that side. On the right side, the assimilation is less perfect, and the bony mass representing the ala is smaller. Ossification of the ligaments between the bones has occurred on the right side, producing complete ankylosis. There is also a partial bony ankylosis between the body of the lumbar vertebra and the anterior surface of the sacrum. The deformity has produced a slight obliquity of the pelvis. The right ilium is more vertical than normal, and the body of the lumbar vertebra is placed somewhat obliquely.

Reduplication of Digits

P. 445. The skeleton of a double manus of a Calf with separate epiphyses to the canon bones and phalanges.

P. 446. The skeleton of a double supernumerary limb said to have grown from the shoulder of a Horse. See *Catalogue of Monsters and Malformed Parts* (1831), No. 283.

Talo-calcanean Bridge

P. 447. The astragalus and os calcis of the right and left foot of the same person. In both there is an osseous bridge between the sustentaculum tali and the astragalus. The condition is congenital.

Notes accompanying Drawing No. 71:

"Anchylosis of the whole surface of the joint where there is a capsular ligament."

"Where there is a capsular ligament there is considerable motion and in proportion to the quantity of motion the part will generally be less disposed to anchylose. Yet in every part of the human body I have seen anchylosis excepting the os humeri with the scapula and the first vertebra with the second."

Hunter included this specimen in the series illustrating anchylosis which he arranged according to the degree of movement in the joint.

Malformation with Defect

P. 448. A right radius. The bone is small and abnormally slender. The normal curve of the shaft is exaggerated and its internal border sharp and prominent. The neck forms an angle with the shaft, with the convexity forwards, and is surmounted by a small irregular nodule of bone representing the head.

Generalised Osteosclerosis

P. 449. The upper part of the right femur of an adult. The bone is abnormally heavy and in the transverse section of the shaft it is seen that, while the diameter of the bone is normal, the medullary canal is reduced in width (1 cm.) by thickening of the cortex. The latter is of dense, ivory-like texture and varies in thickness from 7 to 12 mm. The bony increase appears to be entirely endosteal, without any subperiosteal deposit (diffuse endosteal sclerosis). The upper part of the surface of the shaft and adjacent surface of the neck is superficially ulcerated, probably from suppuration; and at the periphery of the eroded area there is a slight deposit of new bone.

Achondroplasia

P. 450. An achondroplastic Puppy, showing extreme shortening of all four extremities. Since the shortening affects primarily the bones, the overlying soft structures are thrown into well marked folds.

See *Catalogue of Monsters and Malformed Parts* (1831), No. 111.

2. *Atrophy*
P. 451-457

Pressure Atrophy

P. 451. The upper part of the skull of a *giddy* Sheep, in which are several small apertures produced by the pressure of hydatids growing within the brain. The texture of the bone around the apertures is healthy; the diploë is not exposed in the thinning and perforation of the skull.

P. 452. Portion of a sternum, part of the posterior surface of which has been absorbed in consequence of the pressure of an aneurysm. The surface is thinly closed with newly formed compact bone.

P. 453. Part of a femur, in which there is a rarefactive inflammation of the popliteal surface resulting from the pressure of an aneurysm. At the upper part the surface has a worm-eaten appearance; inferiorly, and in some situations among the interstices of the small ulcerated apertures, new bone is irregularly formed.

From a patient who died with sloughing of the foot, in consequence of the pressure of a popliteal aneurysm on the vessels.

The bones of the leg are preserved in No. P. 137.

Everard Home: *Lecture Notes*, 1794-5, vol. i, p. 289. Drawing No. 47. See also Nos. P. 89, P. 90 from the same limb.

Notes accompanying Drawing No. 47:

"The posterior surface of the lower end of the thigh bone where there has been an aneurysm of the popliteal artery which had pressed upon the bone so strongly as to give the irritation for absorption. On the circumference of the part pressed there was a disposition for the ossific process in several parts."

P. 454. The lower end of a femur and the upper end of a tibia, exhibiting the effects of the pressure of a popliteal aneurysm. The popliteal surface is superficially eroded and the thickness of the shaft in this situation is much reduced. The compact tissue of both bones is more porous than normal, as the result of chronic congestion.

P. 455. The lower part of a femur showing the effects of the pressure of a popliteal aneurysm. At the lower part of the popliteal surface there is a shallow depression, with reduction of the thickness of the shaft in this situation. Above the depression the superficial layers of the cortex have been expanded by rarefactive inflammation. The whole of the lower extremity of the bone is abnormally light and atrophic.

P. 456. A right femur, the osseous substance of which is more porous than normal. In its upper third the shaft is bent slightly, the convexity being forwards and outwards.

Senile or Disuse Atrophy

P. 457. The upper part of a right femur. The vascular canals in the wall of the shaft are of large size and the osseous tissue of the anterior part of the neck is light and porous.

3. *Fracture*

P. 458-570

Skull

P. 458. Portion of bone removed, with a trephine, from a boy's skull. It is traversed by a fracture, the edges of which are united by a soft tissue.

Everard Home: *Lecture Notes*, 1794-5, vol. ii, p. 144. Everard Home, On the functions of the brain, *Philosophical Transactions*, vol. civ, p. 484.

FIG. 3
Part of a femur affected by pressure from an aneurism. Drawing No. 47 by William Bell. Specimen 453.

P. 459. A calvaria showing results due to a fracture and trephining.

On the inner aspect there is a long fissured fracture for the most part healed, which passes irregularly forwards from the right side across the whole of the bone: its edges are rounded from absorption, the course of the injury being now traceable as a narrow furrow; at its left extremity, however, the fissure is open, although each of its edges is smoothly rounded and healed.

On the outer aspect, corresponding with the course of the injury described, there is a broad shallow depression in the frontal bone; the surface of this is smoothly healed, and to it there was probably adherent an equally extensive cicatrix, the superficial layer of the calvaria having apparently been destroyed by the growth of granulation tissue following its exposure.

In three positions there are circular apertures due to past trephining; the edges of these are all smoothly rounded by absorption, but they show no attempt toward closure by any growth of new bone.

P. 460. The upper part of a skull on which the trephine had been used twice in the course of the coronal suture. Of the sagittal suture only a small portion remains at its junction with the coronal; and from the end of this portion a fissured fracture extends into the right parietal bone. There appears to have been a depressed fragment at the site of the trephine.

P. 461. Part of a skull which was extensively fissured. The chief force of the injury appears to have fallen upon the upper part of the lambdoid suture; for here a portion of bone, nearly surrounded by fissures, is slightly depressed and starred at its centre. From the fissures surrounding this part, two others extend across the parietal bone to the left temporal bone, and a third, nearly in the course of the obliterated sagittal suture, to the right superciliary ridge of the frontal bone. None of the fissures is closed, but the margins are smoothly rounded. Drawing Nos. 29, 30.

Nose

P. 462. Part of a skull. The ossa nasi have been fractured transversely across the middle of their length, and are reunited, with a slight lateral displacement of the lower portions. These portions are also united by bone to each other and to the superior maxillary bones.

Clavicle

P. 463. A right clavicle fractured through the middle of its shaft, and so completely repaired that there is little trace of the injury.

P. 464. A right clavicle fractured through the middle of its shaft and repaired, with considerable displacement. The scapular portion has passed far under the sternal one; and they are united by a large irregular bridge of bone.

P. 465. A left scapula with the clavicle. The clavicle has been fractured about the middle of the shaft, and the widely separated fragments are united by a flat transverse bridge of bone.

P. 466. A left clavicle fractured through the middle of the shaft and repaired, with considerable displacement. The scapular portion has passed behind the sternal one; and the bridge of bone uniting them is smooth and flat.

Drawing No. 33.

Scapula

P. 467. A right scapula which was split from its base through a part of its infraspinal fossa. The fissure is not closed, but osseous matter is formed about its margin.

P. 468. The scapula of a Lion, of which a part of the base appears to have been broken off. Osseous matter is formed along the margins of the fractures.

Costal Cartilage

P. 469. A portion of a costal cartilage, repaired by bone, after fracture.

The fragments, which overlap for a distance of 2·5 cm. (1 inch), are firmly united by a thick plate of external callus which intervenes between their apposed surfaces. The callus consists of well developed cancellous tissue closed at the surface with a thin layer of compact; over the latter there is a periosteum which is continuous with the perichondrium.

Parkinson's Lecture Notes, 1785 and 1786, vol. ii, p. 72.

Rib

P. 470. Part of a rib "apparently from a Hog" fractured and reunited, with a great accumulation of bone around the injured part. The surface of the new bone, or provisional callus, is marked by an irregular circular groove, indicating that the ossification of the callus is incomplete round the plane of fracture.

P. 471. Portions of three fractured ribs. The broken extremities are slightly displaced and no union has been effected.

P. 472. A second rib, fractured through its middle, and united without displacement The situation of the fracture is indicated only by a rough thickening of the borders and the inner wall.

P. 473. An eleventh rib, fractured through its middle, and united without displacement but with considerable thickening and the formation of ridges of bone on both its borders.

P. 474. Several ribs of a Chameleon, each with one or more small spheroidal osseous swellings upon it.

The altered direction in the axis of the rib, in certain of the specimens, indicates that the swellings represent the external callus formed in the repair of fractures.

"They are, however, more probably scrofulous tubercles, as on some of the ribs there are several of these enlargements in succession,"

MS. Catalogue.

P. 475. Part of the fractured rib of a Cow, of which the broken portions have been partially united by new bone formed on their adjacent margins.

P. 476. The rib of an Ox which was fractured very obliquely and into three or four pieces. Its portions are united imperfectly, the dorsal portions having been drawn forwards from the sternal one; but there is no lateral displacement, and the union is firm.

P. 477. Part of the rib of an Ox fractured transversely. Its portions are united with but little displacement—imperfectly on the outer side, but very firmly and with considerable thickening on the inner wall.

P. 478. The first left rib of an Ostrich fractured near its head, and reunited with considerable lateral displacement of its portions and with a large accumulation of bone.

P. 479. The second left rib of the same Ostrich, similarly fractured and repaired.

P. 480. The third left rib of the same Ostrich, similarly fractured and repaired.

P. 481. The seventh left rib of the same Ostrich, fractured at a greater distance from its head, and reunited with less displacement and less accumulation of bone.

P. 482. The eighth right rib of the same Ostrich, fractured and repaired like the rest.

Humerus

P. 483. Section of the humerus of an Ostrich, fractured transversely through its shaft. The interior of the bone, for nearly an inch above and below the fracture, is filled with a soft tissue, apparently organised exudate; its exterior is unaltered.

P. 484. A longitudinal section of some long bone fractured transversely.

The fractured ends are in exact apposition and firmly held together by an encircling ring of external callus which has undergone ossification. The spaces of the coarse cancellous tissue partially occupying the medullary canal are, at the site of injury, filled in for a short distance with similar callus. The line of fracture across the compact wall of the shaft itself is still distinct and partially open, the intermediate callus not having as yet appeared.

P. 485. A vertical section of a humerus which has been fractured with slight obliquity, near the middle of the shaft, and has united with but little displacement. The specimen has been minutely injected and afterwards decalcified, and is now preserved in alcohol.

The medullary canal has been restored at the site of the fracture, and all the redundant portions of external callus removed.

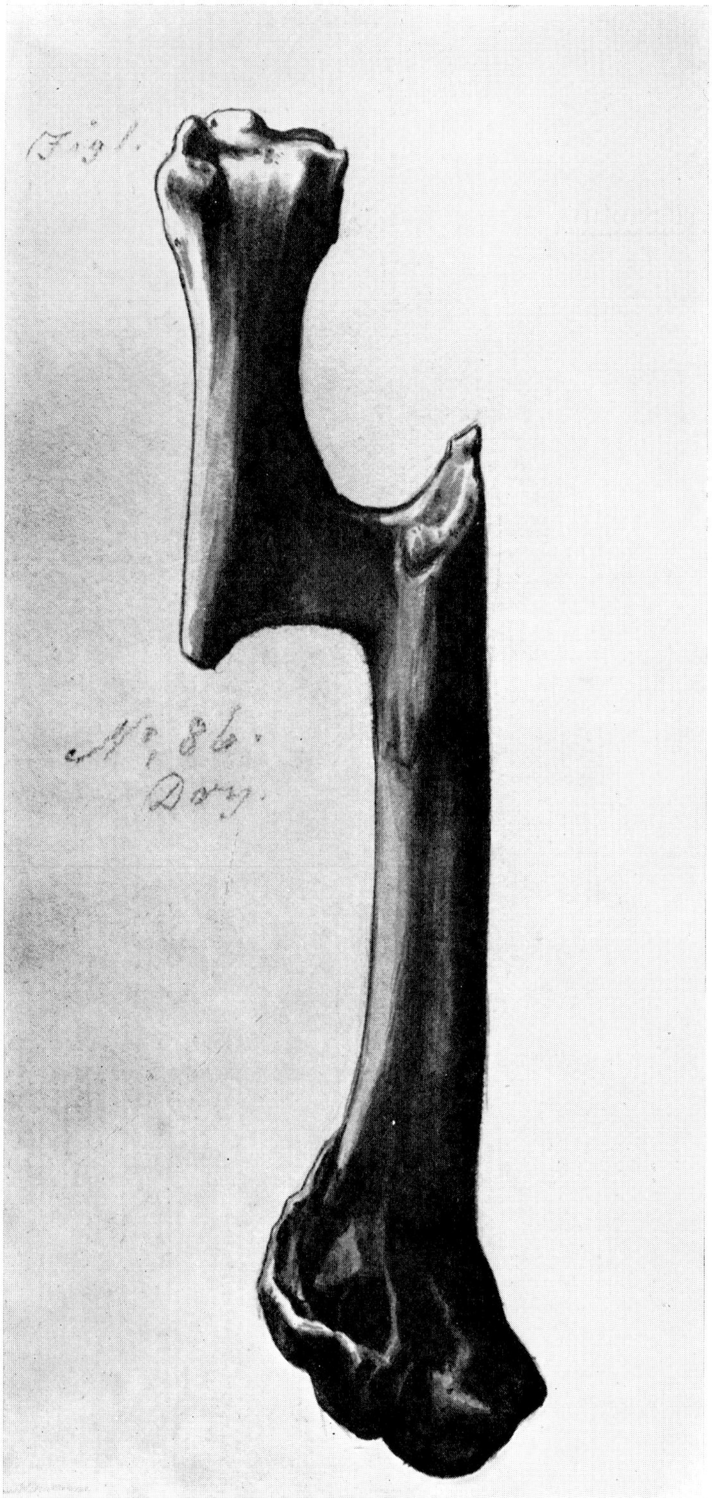

Fig. 4

"The Os Humeri of a Turkey, where the two broken ends of the Bone were about an inch from each other, probably owing to the hanging of the wing."
Drawing No. 34 by William Bell. Specimen 487.

Face p. 13

P. 486. The humerus of a Vulture fractured just below the middle of its shaft. The extremities of the portions were displaced nearly an inch from each other, but have been united by a strong, smooth bridge of bone.

P. 487. Longitudinal sections of the humerus of a Turkey which has been fractured near the middle.

The fragments have undergone longitudinal displacement so as to overlap for a length of 2 cm. ($\frac{3}{4}$ inch), and are, in addition, widely separated laterally for an intervening distance of 1·3 cm. ($\frac{1}{2}$ inch). Firm union has, nevertheless, been brought about by means of a flattened bridge of external callus produced between the surfaces which face one another.

The tissue of the callus is uniformly cancellated, except at the surface where it is formed of compact tissue continuous with the walls of the shaft. The pneumatic canal of each fragment is closed with internal callus.

Drawing No. 34.

P. 488. A right humerus, the shaft of which was broken in two places about the junction of its middle and lower thirds. The portions are united firmly and evenly, but with a considerable displacement of the lower portion, which is fixed anteriorly and externally to the axis of the shaft. A smooth depression exists where the medullary cavity of this lower portion is covered in with a thin plate of bone.

Drawing No. 31, fig. 2.

P. 489. The humerus of a young Ostrich in which a fracture has been repaired after slight angular displacement.

The pneumatic canal has been completely restored by the removal of the internal callus, and the redundant external callus has been absorbed. The compact tissue of the fragments is indistinguishably continuous across the line of fracture.

P. 490. A left scapula, clavicle, and humerus. The great tuberosity of the humerus, with an adjacent portion of the border of the head, has been broken off. It is firmly reunited to the shaft, though but partially, for only one border of it is fixed and the rest is widely separated. Below it some small splinters seem to have been broken from the shaft and reunited. There is some superficial erosion and new bone formation round the glenoid cavity.

P. 491. A right humerus which was fractured transversely just below its tubercles. The upper end of the shaft appears to have been drawn forwards, and is united by a short bridge of bone to the anterior border of the head and neck. The head and its tubercles are reduced to a disc-like mass, less than an inch in thickness, and having externally a smooth flat surface continuous with the outer wall of the shaft. The articular surface of the head is ulcerated at its centre; and bone is irregularly accumulated around its margin.

P. 492. A right humerus, fractured and repaired in almost exactly the same manner as that last described. It was also fractured obliquely through the lower third of its shaft; and these portions are united with a slight lateral displacement.

P. 493. A left humerus fractured obliquely just below the head and the tuberosities. The portions are firmly united; but the shaft is drawn inwards and a little forwards, so that it is placed almost directly under the head, and the prominence of the tuberosities is greatly diminished.

P. 494. A nearly similar specimen, with a large ridge of bone formed on the outer margin of the bicipital groove.

P. 495. A right humerus fractured obliquely through the middle of its shaft. The lower portion has been drawn upwards and backwards, but firm reunion has been effected by intermediate new bone.

P. 496. A left humerus fractured rather obliquely through the middle of the shaft. The parts are united firmly, but with some lateral displacement. A large fragment, which was separated from the posterior part of the shaft, is now firmly fixed like a bridge passing from one part of the shaft to the other.

P. 497. A right humerus fractured transversely through the junction of the middle and lower thirds of its shaft. The portions are united firmly; but the lower lies almost entirely in front of the upper portion. The parts of the medullary canal, which were thus exposed at the end of each portion, are closed with bone; but slight depressions indicate their former situation. The extremities of the portions are smoothly rounded; but a small osseous spine projects from the margin of each.

P. 498. The humerus of a Hare, which was fractured near the middle of its shaft. The portions overlap each other considerably and are not united, though a large quantity of new bone is formed upon their adjacent surfaces.

P. 499. The humerus of a Turkey fractured in two or more directions near the middle of its shaft. The portions are united firmly, but with such displacement that they form nearly a right angle with each other.

P. 500. The humerus of a Fowl fractured about the middle of its shaft. The parts are much displaced, but are united by an intermediate broad piece of bone.

P. 501. The humerus of an Eagle fractured in two places near the middle of its shaft. The two lower portions are united with much displacement, the lowest of them having been turned half round; the upper portions are not united, but the broken end of the medullary cavity is closed and new bone is formed around its surface.

Ulna

P. 502. Section of the ulna of a Fowl fractured obliquely through the distal part of its shaft. The broken extremities are rather widely separated, but are held together by a firmly attached ring of ossifying callus, irregular in its form and from half an inch to an inch in width. The interior of the bone near the seat of fracture is filled with a callus, but its walls appear to have undergone no change.

"The ulna of a Fowl where the callus has become cartilaginous, which is to be seen in those places that are not yet ossified."

Note by Clift: This earlier description seems better to express Hunter's meaning, that the callus had become cartilaginous preparatory to ossification. *MS. Catalogue.*

P. 503. The upper part of an ulna in which there is extensive erosion of the articular surfaces, especially in the lesser sigmoid cavity. A fracture, passing vertically downwards from the joint through the base of the olecranon, is firmly united by interposed bone.

Radius

P. 504. A vertical section of a radius in which there was a compound fracture. The two portions are wide apart, but are united by a large quantity of tough ligamentous tissue containing a cavity in which was a small detached portion of bone.

P. 505. The other section of the same radius, together with the detached piece of bone between the ends of the fractured portions.

P. 506. A left radius fractured rather obliquely at the middle of its shaft. The portions are united, with a slight displacement.

P. 507. A right radius similarly fractured, but in which the apposition of the portions was less exact, the lower portion having passed upwards in front of the upper portion.

P. 508. The bones of a forearm, with the carpus and metacarpus. The radius has been fractured about three fourths of an inch from its carpal end. The portions are united firmly and smoothly; but the lower portion has been driven backwards and towards the radial side of the upper portion. The radius is shortened half an inch; and, the ulna retaining its natural length, the carpus and hand are in consequence inclined to the radial side, and the ulna appears unnaturally prominent. There is an appearance, also, as if the styloid process of the ulna had been broken off and reunited in its natural position.

Metacarpus

P. 509. Section of a metacarpal bone in which there had been a compound fracture. The portions are widely displaced, but are united by a soft callus.

P. 510. The other section of the same metacarpal bone.

Pelvis

P. 511. The pelvis of a Deer, in which the ossa pubis and the left ischium were fractured in several directions. Their portions are firmly reunited, with considerable accumulation of bone and but little distortion.

Tail

P. 512. The bones of the tail of an animal fractured in two places and united with some displacement of the fragments.

Femur

P. 513. Section of the upper part of a femur and of the acetabulum. There has been a fracture directly through the neck of the femur near its junction with the head. All that portion of the neck which intervened between the plane of fracture and the shaft has been absorbed; and the head, unaltered in its form and structure, is attached superiorly to the neck by remains of the overlying reflection of fibrous tissue. Everard Home: *Lecture Notes*, 1794-5, vol. i, p. 138.

P. 514. The other section of the same femur, showing more distinctly than the preceding specimen, a thickening and contraction of the capsular ligament.

P. 515. The lower part of a femur, fractured obliquely at the junction of the middle and lower thirds.

The upper fragment has been displaced downwards, behind, and to the inner side of, the lower. The ends of the bones are enclosed in an irregular cavity bounded by the muscles and connective tissue, and lined, like the fractured surfaces, with blood clot.

A quill has been passed through the popliteal artery, which has been ruptured in the injury.

P. 516. Transverse sections of the left femur of a Fowl which was fractured near the middle of its shaft. The lower portion was drawn backwards and more than $\frac{1}{2}$ inch upwards, and was turned half round; but it has been firmly reunited by an intermediate mass of external callus. The ends of the medullary tube exposed by the fracture are closed by smooth hard layers of bone.

P. 517. A right femur fractured through the upper part of its shaft. The lower fragment is displaced above and behind the upper, for a distance of about 2 inches: union has taken place by the formation of two lateral bridges of external bone.

P. 518. The femur of a Fowl which was fractured in two places below the middle of its shaft. The portions have united firmly, the lower portion being displaced backwards so as to form an acute angle with the upper.

"It may be relative that the bones of birds are almost always united by luxuriant callus, and the reason is obviously on account of their greater activity or restlessness when compared with quadrupeds." *MS. Catalogue.*

P. 519. A section of the femur of a Cock in which, after it was broken into three pieces, the upper portion was widely displaced. The two lower portions have firmly united by bone, and the space between them and the upper portion is filled by a tough fibrous-looking tissue, by which they are securely held together, and in which there is a small isolated portion of newly formed bone.

P. 520. The other section of the same femur.

P. 521. A left femur showing the repair of two fractures.

At the upper end the neck with the posterior and greater portion of the trochanter major has been detached from the shaft. Osseous union has occured after some displacement from the lower fragment upwards in front of the upper.

The second fracture passes very obliquely from without downwards and inwards through the shaft about the junction of its middle and lower thirds. The lower fragment has been displaced upwards and to the inner side of the upper. In addition to this, a splinter, 11·5 cm ($4\frac{1}{2}$ inches) in length and 2 cm. ($\frac{3}{4}$ inch) in breadth, has been split from the inner side of the lower fragment, "the splinter having by some extraordinary circumstance been turned round on its bone, so that it must have been quite detached, its periosteal surface is laid on that of the bone, and the internal or reticular surface presented to the muscles; but what is extraordinary is, the strong bony attachment by which it is fastened to the original bone, when, from the change it had suffered in its situation, it must have been entirely detached".

A fair amount of osseous union has occurred at the site of the lower fracture. The splinter is firmly united with the rest of the femur, and a certain amount of finely porous new bone has been produced on its medullary aspect. Although the splinter in question was completely detached from the rest of the

FIG. 5

"A Thigh Bone where there has been two fractures, one in the neck which is pretty well united; the other in the body of the Bone in which there is a singularity; the Bone for some way has been broken into splinters, one of which must have been fairly detached from all the surrounding parts and turned so far round as the canal to become external." Drawing No. 37 by William Bell. Specimen 521.

Face p. 18

bone, it doubtless maintained some connection with portions of the vascular structures attached to its periosteal surface.

Parkinson's Reminiscences, 1833, p. 111. Everard Home: *Lecture Notes*, 1794-5, vol. i, p. 147. Drawing No. 37.

P. 522. The upper part of a right femur which has been fractured through the higher end of the shaft, the great trochanter being in addition comminuted, and the lesser trochanter detached. The various fragments have undergone considerable displacement.

An unusually large amount of external callus has been formed about the site of injury, particularly along the lower and posterior aspects.

P. 523. The femur of a young adult in which a transverse, compound fracture has occurred at the junction of its middle and lower thirds. The lower fragment has been displaced upwards behind, and to the outer side of the upper for a distance of 7·5 cm. (3 inches).

For about 5 cm. (2 inches) the end of the upper fragment has undergone total necrosis; the dead portion, which is probably conical in form and extends some way further up the shaft, is in process of detachment. The edges of the fracture are perfectly sharp and unaltered, and the surface of the dead piece is quite smooth. An extensive amount of new bone has been produced on the shaft above the exposed part of the sequestrum.

Of the end of the lower fragment, only small portions have undergone necrosis. The medullary canal is filled with internal callus; bridges of the external pass between and unite the laterally separated living ends of the two fragments.

P. 524. The right femur of an adult, which exhibits the results due to past rickets, the shaft being curved unnaturally forwards, and the neck of the bone sunk to a right angle.

The shaft, at the junction of the upper and middle thirds has been the seat of a very oblique fracture which has united after displacement of the lower fragment upwards in front of the superior. The fragments are firmly united by intervening callus, except in two situations where open intervals exist.

P. 525. A left femur, of which the middle third of the shaft has been broken into many large pieces.

The several fragments are firmly reunited after some shortening and distortion. The medullary canal of the highest fragment, which projects in advance of the rest of the bone, is closed with internal callus.

Drawing No. 36, fig. 1.

P. 526. The lower part of a femur which was fractured very obliquely from above downwards and forwards. The upper fragment has been driven more than three inches downwards in front of the lower; and its sharp end must have penetrated the knee-joint for it is now on a level with the condyles.

The pointed end, with a long strip of the inner border of the upper fragment, has undergone necrosis, the dead portion being surrounded by a shallow jagged groove of demarcation. There is a similar area of necrosis at one spot on the inner border of the lower fragment, which is vertically split. A scanty formation of new bone is to be seen here and there on the surface of the lower fragment: the gap in the outer border of this fragment was evidently existent during life.

P. 527. The right femur of a Sheep, which has been fractured near the middle of its shaft. Union has occurred after considerable longitudinal displacement, the lower fragment being drawn upwards in front of the higher.

An excessive production of external callus has taken place at both ends of the union, but particularly at the lower, where flattened processes have extended for some distance into the surrounding soft parts.

P. 528. The upper end of a femur. Below the situation of the trochanters the shaft exhibits an angular forward bend, without other distortion. The medullary cavity is filled with cancellous tissue at the site of the curvature; and on the posterior aspect of the bone the surface is trabeculated from a formation of new tissue.

The curvature of the shaft is probably the result of an old incomplete fracture, in which the convex wall has been broken and the concave only bent. The head and neck of the bone are quite normal, and present no trace of changes due to past rickets.

P. 529. Part of a femur, in which a compound fracture occurred just above the condyles. A portion of the cancellous tissue underwent necrosis and is contained in a large cavity between and behind the ends of the bone. The cavity has vascular walls and an external opening through a fistulous passage above the inner condyle. The ends of the fractured bone are rounded and smooth; but no process of union has taken place. *Parkinson's Lecture Notes*, 1785, vol. ii, p. 62.

P. 530. The upper part of the femur of a young person, in which necrosis followed fracture. The dead bone, comprising about 1 inch of the fractured end of the femur, is surrounded by a broad, shallow, ulcerated groove. A part of the shaft above this groove has a thin deposit of new bone upon its surface. It consists of short and very fine osseous fibrils and lamellae, which are rolled and curled so as to enclose minute cellular and tubular spaces and channels. These are closely arranged and form a dry, friable tissue, with a very finely grooved and perforated surface, and with an indistinct appearance of longitudinal *grain* derived from the direction of the chief lamellae. The new bone is closely connected to the surface of the subjacent old bone; but it may be scraped off; and the surface of the bone thus exposed (as shown at the sides of the specimen) appears only a little roughened by the enlargement of the vascular grooves on its surface.

P. 531. A right femur, of which the shaft was broken near the junction of the middle and lower thirds. The portions considerably overlapped each other, the lower one having passed far upwards behind the upper one; but they are united by a strong intermediate mass of bone and, though there is great shortening of the limb, there is no eversion of the condyles. Drawing No. 36, fig. 2.

P. 532. A left femur of which the shaft was fractured just above its middle. The portions were widely separated, the upper one having been drawn forwards and abducted, and the lower upwards and backwards behind the upper; they are firmly united by a strong intermediate mass of bone.

P. 533. Part of the lower main fragment of a fractured femur divided vertically. The fracture has passed almost transversely through the shaft and the fractured extremity has undergone necrosis. There is a well marked groove of demarcation bounding the dead bone, and the shaft for 3 inches below is partly covered with a thin layer of new bone. Drawing No. 10.

P. 534. Part of a femur in which, after fracture through the middle of the shaft, necrosis of the ends of the fragments took place. The dead portions were in process of separation, and new bone was abundantly produced on all the parts around them; but no union has been effected.

The bone of the shaft is abnormally porous as a result of suppuration.

P. 535. A right femur which fractured obliquely through the upper part of its shaft. The portions have failed to unite and part of the inner surface of the lower fragment has suffered necrosis; and there has been an abundant accumulation of bone around its margins, as well as on the adjacent part of the upper portion.

P. 536. A left femur fractured with comminution through the upper part of its shaft, and the base of the great trochanter. There appears to be no osseous union. (The portions are artificially united.) The bone is abnormally porous in the neighbourhood of the fractures and for some distance around, due probably to suppuration.

P. 537. A left femur fractured through its shaft, just below the lesser trochanter. Its portions are united smoothly and firmly, but with some shortening and a prominent angle in front. The shape of the head and neck is altered by chronic osteoarthritis. Drawing No. 35.

"When a thigh is broken it is generally shortened. A large bump shall make its appearance after the swelling has subsided, which was not observed before, and this often gives an alarm to the patient that the bone is not united, or that it is ill-set. This is the case more frequently in the thigh as the ends of the bone generally ride over each other more than any other because the muscles of the thigh are strong and unruly, that the bone is seldom got, and kept, in its proper situation. Therefore in cases of broken thigh great attention should be paid to prevent as much as lies in our power its shortening." Everard Home: *Lecture Notes*, vol. i, p. 157.

Tibia and Fibula

P. 538. Part of the shaft of a tibia from a case of compound fracture at St. George's Hospital. The periosteum has been separated from a large area of the bone: the membrane is thickened and its inner surface roughened. At its fractured end the whole thickness of the shaft has undergone necrosis, the dead bone being marked off from the adjacent bone by a distinct groove. See also P. 551. The fibula from the same case.

P. 539. A vertical section of a fractured tibia. The continuity and form of the wall of the bone are completely restored. The parts have been injected and decalcified.

P. 540. A vertical section of the lower end of a fractured tibia, made after the parts had been injected, and softened in acid. There was considerable displacement of the portions of bone, and only the halves of their fractured surfaces have united. To this extent, however, the union is complete, and the line of fracture across the cancellous tissue is scarcely discernible.

"From a case of compound fracture at St George's Hospital. The sore (consequent on the protrusion of the bone) never would heal; nor could the man ever bear any weight upon it; probably from its having united by so small a surface; the limb was therefore amputated."

MS. Catalogue. Parkinson's Lecture Notes, vol. ii, p. 49.

P. 541. Another section of the tibia from the same case. The tibia has been fractured transversely about 1 inch above its lower extremity and the lower fragment has been displaced laterally so as to lose all contact with the upper fragment and the two are connected merely by a narrow plate of new bone. The cancellous tissue of both fragments exposed by the fracture is closed by osseous tissue. The whole shaft appears to have been unnaturally vascular.

P. 542. Portion of the shaft of a tibia which has been obliquely fractured with comminution, a somewhat quadrilateral splinter having been widely displaced amongst the muscles over the outer aspect of the two main fragments.

A finely porous bony callus is in process of formation within the medullary canal, and on the exterior of the upper fragment.

P. 543. Section of a fractured tibia. The portions of the bone are rather widely separated, and the space between them is filled with a soft vascular substance.

P. 544. A vertical section of the lower end of a tibia which is the seat of a comminuted fracture, and of which one of the fissures passes into the ankle joint.

Black bristles have been used to mark a narrow layer of new tissue, or of lymph, filling the line of one of the fissures. A thin layer of reparative tissue invests the irregular excavated surface beyond, which has been exposed, apparently, by the removal of the fragment which lay in it.

A blood-stained layer of coagulum, or lymph, has been peeled from off the articular cartilage.

P. 545. The other section of the same tibia, showing similar changes, with a somewhat wider separation of the fragments.

P. 546. A longitudinal section of portion of the shaft of a tibia, in which an oblique fracture has been repaired after slight longitudinal displacement, the posterior wall of one fragment lying in the axis of the anterior wall of the other.

The medullary canal of each fragment is filled with internal callus, and callus has been produced externally in such a way as to fill the angles resulting from the overriding of the fragments. The most exterior portion of the callus is as dense as the compact walls themselves. The sharp edges of the fragments have been smoothly rounded by absorption.

P. 547. A vertical section of a tibia the whole of which, with the exception of the upper articular end, has been the seat of chronic inflammation, probably from syphilitic infection.

Near the junction of its upper and middle thirds the shaft has been obliquely fractured, the upper fragment being displaced slightly in advance of the lower. Complete osseous union has taken place.

(The specimen may illustrate the occurrence of a fracture in a bone previously inflamed; or the occurrence of syphilitic inflammation in a bone in which a fracture has been previously repaired.)

P. 548. A tibia after a compound fracture through the middle of its shaft. The portions are firmly united, but with lateral displacement. The new bone is abundant and heavy; there is a cavity in its interior; and its surface is rough and porous like that of new bone formed on bones inflamed. There is also a cavity between the ends of the united fragments, which probably opened externally through ulcerated integuments.

P. 549. The adjacent portions of the fragments of a fractured tibia.

As a result of infection (the injury having doubtless been compound) parts of the ends of the fragments have undergone necrosis, and are in process of detachment. A small amount of new bone has been produced on the living parts of the shaft beyond those which have died.

P. 550. A fibula very obliquely fractured through the lower part of its shaft and united with but little displacement.

P. 551. Part of the fibula of a man who had a compound fracture of both the bones of the leg. The upper fragment is nearly an inch below the lower. Its fractured surface is almost smooth, the cancellous tissue being filled by callus; and it rests in a cavity formed by a buttress of external callus which projects from the other fragment. See P. 538. Tibia from the same case.

P. 552. Part of a tibia with the adjacent tissues, from a case of compound fracture. Portions of the bone have been removed; and the remaining fractured ends, an inch and a half apart, project into an extensive granulating cavity.

Everard Home: *Lecture Notes*, 1794-5, vol. i, p. 139.

P. 553. A tibia and fibula fractured obliquely about an inch above the ankle-joint. Some blood is effused and coagulated between the broken surfaces. There is a small aperture in the skin, through which the upper portion of the tibia protrudes.

P. 554. The lower ends of a tibia and fibula. There is a comminuted fracture of the tibia, involving the ankle-joint. The fissures are occupied by soft organised tissue. Parts of the articular cartilage of the tibia and fibula have been removed by ulceration.

P. 555. A left tibia and fibula. There is a comminuted fracture involving the upper extremity of the tibia and the upper fourth of the shaft, several of the fissures extending to the articular surface. There is no evidence of repair. (The fracture of the fibula is artificial.)

P. 556. A left tibia and fibula, of which the former is fractured with comminution in the lower third of the shaft, the latter between 2 and 3 inches below the knee. A thin splinter has been broken off the back of the tibia; and a fissure extends into the ankle-joint. No process of repair appears to have commenced.

P. 557. Part of a tibia which was fractured near the middle of its shaft. The fragments have united with marked angular deformity. New bone has been formed on the surfaces of the extremities of both fragments and, especially, in the angle behind them; at the anterior part of the angle one of the bones is rather deeply ulcerated.

The fracture was probably compound.

P. 558. Section of part of a tibia fractured with comminution near the junction of the middle and lower thirds of its shaft. The portions are not united, but new bone has been formed upon their outer surfaces.

From a case of compound fracture.

P. 559. Sections of one of the main fragments of a fractured tibia in which a portion of the fractured extremity suffered necrosis. The dead bone is abnormally porous, owing to inflammation preceding its necrosis, and there is an abundant deposit of new bone on the living bone above it.

P. 560. The lower end of a right tibia which has been fractured about $2\frac{1}{2}$ inches above the ankle-joint. The fractured end of the bone has undergone necrosis and the adjacent parts are covered with an abundant callus.

P. 561. The tibia and fibula of a Fowl. The tibia was fractured obliquely through the middle of its shaft: its portions are united with but little displacement or accumulation of bone. The fibula, of which only the lower part remains, opposite the fracture is enlarged and fixed to the tibia by a superficial formation of new bone.

P. 562. The tibia of a Fowl which was fractured almost transversely below the middle of its shaft. The portions overlap to a considerable extent, but are united firmly and with but little accumulation of bone. The ends of the medullary tube are closed.

P. 563. A right tibia and fibula which have been fractured with comminution, the former in its lower third, the latter in its upper third. The bones have united firmly with considerable angular deformity. Firm union has taken place, but with considerable distortion and superficial formation of new bone. The lower extremity of the upper fragment of the tibia forms a sharp projection at the inner part of the bone.

P. 564. Parts of a tibia and fibula which have been fractured just below the middle of their shafts. There is extreme angular deformity, the fragments being nearly at right angles. The fibula has united by bone, the fragments overlapping. No union has occurred in the tibia; the extremities of the fragments have become rounded off and there is a slight deposit of new bone; the adjacent parts of the shaft are pitted and eroded by inflammation.

From a case of simple fracture.

"The surface of the two broken bones had become smooth, fitting them for motion: this probably arose in some measure from want of attention in keeping the bones at rest, for they are nearly at right angles which shows the want of chirurgical assistance. The fibula, which has evidently been in the same condition that the tibia still retains, had become united and served the purpose of a splint to the tibia, in which some slight efforts to reunion had begun to take place." *MS Catalogue.* Drawing No. 38.

P. 565. A left tibia and fibula, of which the former has been obliquely fractured between 3 and 4 inches from the ankle, the latter between 2 and 3 inches from the knee. Both fractures are firmly united with some longitudinal displacement. The lower extremity of the upper fragment of the tibia forms a prominent projection on the inner aspect of the bone.

P. 566. A tibia and fibula. The tibia has been fractured near its lower end. There has been considerable loss of substance at the seat of fracture and a sequestrum about 4 inches in length has separated from the anterior aspect of the upper fragment. An abundant formation of new bone has occurred at the site of the necrosis and on the lower fragment, which at its upper part has become ankylosed to the fibula. The fibula has been fractured near its upper end, and is firmly united by bone.

P. 567. A tibia and fibula which have been fractured about 2 inches above the ankle-joint. The osseous tissue of the extremity of the upper fragment of the tibia is smooth and unaltered in appearance and has undergone immediate necrosis; the rest of the shaft in almost its whole length is irregular and eroded from inflammation and has undergone necrosis at a later stage. On the outer surface the distinction between the living and dead bone is well demarcated; on the inner and posterior surfaces the distinction is less marked. The lower end of the upper fragment of the fibula is firmly fixed to the lower fragment of the tibia, and a considerable quantity of new bone has been formed about them. Drawing No. 81.

P. 568. Portions of a fibula in which, after a compound fracture through the lower part of its shaft, necrosis of parts of both the fractured ends ensued. The dead bone is separated from the living by a shallow groove. New bone has been abundantly deposited upon both portions of the shaft.

P. 569. The extremity of one of the main fragments of a fractured tibia. An irregular portion of the cortex of the fractured end, including nearly the whole circumference of the shaft, has undergone necrosis and has almost completely separated from the living bone. The latter is abnormally porous and at one part has a thick deposit of new bone.

Foot

P. 570. The metatarsal bone of a Sheep, of which a great part of both the walls and the cancellous tissue has been destroyed by ulceration or necrosis, or both, succeeding to a fracture through the middle of its shaft. The fractured portions are united, with considerable displacement. *Parkinson's Lecture Notes,* vol. ii, p. 79.

4. *Non-union and False Joints*
P. 571-575

Tibia

P. 571. A vertical section of a tibia fractured with slight obliquity through the lower part of its shaft. An irregular ferrule of external callus has been produced around the ends of the fragments; but no union has taken place elsewhere. A portion of the posterior wall of the higher fragment is necrosed and detached.

Humerus

P. 572. The right humerus of an Ostrich which was fractured just above the lower third of its shaft. The portions are not united, but the end of each is enlarged, by the accumulation of new bone around it, into a smooth oval mass of compact bone of more than twice the natural diameter of the shaft.

P. 573. Sections of the humerus of a rickety Monkey of which the shaft has been fractured; the fragments which lie at an angle have remained ununited, with formation of a false joint. The shaft is greatly thickened by the formation of new spongy bone, which entirely fills the medullary canal. The extension of bone into the cartilage at the upper end is very irregular.

Femur

P. 574. A transverse section of a fractured femur in which the portions are united by external callus, one lying upon the other.

Ulna

P. 575. A right ulna which has been transversely fractured through its middle, and between the fragments of which a false joint has formed.

Each of the fragments is lipped with external callus; the medullary canal at the site of fracture is filled with internal callus, and the broken surface of the compact wall is covered with intermediate. No union has occurred, however; the apposed and articulating surfaces of the callus of the two fragments being smoothly closed with a thin layer of compact bone. The lower fragment is displaced slightly inwards at an angle with the upper.

5. *Amputation Stumps*
P. 576-606

Disuse Atrophy

P. 576. A section of the stump of a tibia. The bone is reduced in size, tapering towards the extremity. The medullary canal is closed with compact bone, and the extremity of the stump is sharply recurved.

P. 577. The upper part of a tibia and fibula from an amputation stump. The bones are atrophic, the medullary canals are closed by a deposit of new bone on their extremities, and the ends of the two bones are ankylosed.

P. 578. The stump of a fibula, after amputation near the knee-joint, reduced in size, and tapering to its extremity.

P. 579. The stump of a tibia, after amputation very near the knee-joint. The end of the stump is much reduced in size, and its anterior surface is covered with new bone, which is very light and cancellous near the end, but more compact and laminated near the knee-joint.

Fusion of Bones

P. 580. Part of a right tibio-tarsus of a Turkey, the end of which has been lost by some injury.

The extremity of the bone is covered with a pitted layer of new osseous tissue. The ossified tendons of the flexor muscles are united to the end of the bone.

Bony Spurs

P. 581. The stump of the humerus of an Eagle. From the opened pneumatic canal there projects a prominent hemispherical mass of porous bone, produced as a result of inflammation affecting the stump.

P. 582. The stump of a femur, of which the lower third has been removed by amputation.

The medullary canal is closed with new bone. The sawn edges of the compact wall have been smoothly rounded by absorption; and there is immediately beyond them a sharp, recurved, and spiky process of new osseous tissue which grew into the soft parts adherent to the end of the femur. Drawing No. 79.

Periostitis

P. 583. The end of a tibia after amputation. A ring of the sawn end has undergone necrosis and is in process of separation. A layer of new bone is formed over the surface of the living bone above a broad ulcerated zone of demarcation.

P. 584. Portion of the shaft of a tibia, of which the lower end is that divided in an amputation stump.

A large portion of the bone has undergone necrosis, its extent being recognisable by the complete absence of any new formation upon its exterior. The dead piece is surrounded by a deep groove of demarcation along which its separation is proceeding. The living strip of the compact wall is covered with a thick layer of inflammatory new bone.

The upper section has been carried through part of the sequestrum, but whether a second amputation was performed is not recorded.

Separation of Periosteum

P. 585. "A section of a femur which was amputated, with the periosteum separated, to show how clear the periosteum separates from the bone in inflammation; a fact almost always observable after amputation." *MS. Catalogue.*

The separation of the periosteum from the end of the bone indicates that the latter is necrosed. The surface of the bared bone is quite unaltered, it has not even the thinnest layer of new osseous tissue over it.

P. 586. A longitudinal section of portion of the shaft of a femur from an amputation stump.

The end of the bone is bare and necrosed. Above the dead portion, the living is encircled with a low nodular formation of new bone with which the periosteum is incorporated. On the living surface of the shaft, from which the soft structures have been artificially stripped, there is an extremely thin layer of new osseous tissue, which is wanting on the dead piece.

P. 587. Part of the shaft of a femur from an amputation stump. The end of the bone is bare and necrosed. On the living portion of the shaft, there is an extremely thin layer of new osseous tissue, which is absent on the dead portion.

Ring Sequestrum

P. 588. A vertical section of portion of the shaft of a femur, a fortnight after amputation.

The sawn surface has undergone complete necrosis, the extent of which is indicated on the exterior by a groove of demarcation, and in the divided surface by the rarefaction of the living shaft immediately beyond the dead.

The living bone is encircled with a finely porous ring of new osseous tissue which thins off and becomes lost as it is traced upwards on the surface. At the back of the specimen, a portion of this ring has been cut away so as to fully expose the groove of demarcation already referred to. From the tissue occupying the medullary canal there has grown out an overhanging mushroom-like formation of connective tissue: a minute, whiter island of new bone is discernible in the centre of its base of attachment. The medulla within the end of the bone has been replaced by connective tissue like that which protrudes at the surface.

"A piece of the thigh bone of a young lad who lived about a fortnight after the thigh was amputated. The fungus at the end of the bone is seen to come from the marrow. In the marrow near the end of the stump was bony matter forming, and all round the end of the stump was an increased bony thickness; and exfoliation is just begun." *MS. Catalogue.*

P. 589. A vertical section of the end of a femur, after amputation.

A ring of the bone at the sawn extremity has undergone necrosis and is in process of detachment. On the living shaft above, a somewhat voluminous inflammatory formation of finely cancellated osseous tissue has taken place: a small amount of similar bone has been produced in the most distal part of the medulla. The medullary tissue and the new bone on the exterior are covered with granulations.

The blackness of the dead piece is the result of charring, the actual cautery having been applied to accelerate its detachment.

P. 590. The other section of the preceding specimen (P. 589), macerated and dried. The new bone on the shaft, where covered in the living state with granulation tissue, has an open finely cancellous surface.

P. 591. Portion of a fine ring of dead bone which was naturally detached from the sawn end of an amputation stump.

It shows the irregular, jagged surface along which the separation has been effected; the surface actually divided has undergone no change.

P. 592. A section through an amputation stump "to show the appearance of an ulcer over a bone which is exfoliating".

P. 593. Part of the shaft of a femur, in which, after amputation, a narrow ring of the sawn end, comprising its entire thickness, has undergone necrosis.

At the surface the dead portion is marked off by a groove of demarcation resulting from the absorption of the bone by the new formed granulation tissue. The Haversian grooves and canals in the surface of the living bone for some distance beyond are unnaturally enlarged from the growth of granulation tissue about the vessels; and small, thin, patches of very finely porous new bone have been formed on the rarefied surface.

P. 594. "An internal exfoliation from Mr. Maitland's Os Humeri, 1775." *MS. Catalogue.*

Massive Necrosis

P. 595. The upper half of a femur from an amputation stump.

A narrow unaltered ring of the bone at the sawn end has undergone necrosis. For a considerable distance above, the surface of the shaft is eroded. The eroded portions are likewise necrosed, the erosion having been due to the removal by granulation tissue of the most superficial layers of the shaft, which had not themselves died, and the whole process having taken place before the occurrence of suppuration. The line of demarcation at the upper limit of the necrosis is not yet declared. At the higher part of the lesion, the removal of the surface and the subsequent formation of the pits can be traced in an enlargement of the Haversian canals by the growing granulation tissue.

P. 596. Part of a femur, from the end of which an annular portion exfoliated after amputation. The extremity is smooth but porous, like ulcerated cancellous tissue: there is a cavity in its centre, and it is surrounded with a small quantity of new bone.

P. 597. Part of a femur, from the end of which portions exfoliated after amputation. The necrosis seems to have affected only parts of the wall, for the cancellous tissue remains entire in the midst of the cavities from which the sequestra were removed. New bone has been formed abundantly on all the remaining part of the shaft.

P. 598. Part of a femur from an amputation stump. A ring of bone has necrosed and immediately above a large quantity of new bone has been formed.

P. 599. Portion of a tibia, from one end of which (apparently after amputation) a considerable exfoliation was taking place. The adjacent part of the shaft has new bone abundantly formed on its surface.

P. 600. The upper half of a tibia from an amputation stump. At the divided end a piece of the outer wall, about 3·3 cm. ($1\frac{1}{4}$ inches) in length, has undergone necrosis for its whole thickness; and, continued upwards from this as high as the tubercle, a strip of the bone, comprising the crest and outer wall, is likewise dead; the surface of this portion is eroded from the removal of the superficial layers, which have escaped death. On the inner border the sequestrum is surrounded by a deep groove of demarcation; on the outer, there is a similar groove separating the living from the dead portions, but here the groove is limited to the upper area, the parts being elsewhere still uniformly continuous.

New bone has been produced in varying amount on the living portions of the shaft, and within the medullary canal. From the gap in the lower end of the dead piece, a second sequestrum has been detached.

P. 601. The stump of the proximal phalanx of an amputated finger.

Necrosis of the terminal part of the bone has occurred on its dorsal aspect, above which a thin layer of porous new bone has been deposited.

P. 602. A sequestrum, 14 cm. ($5\frac{1}{2}$ inches) in length, from the shaft of a femur after amputation.

At the sawn extremity the whole thickness of the shaft is involved; higher up the surface is almost everywhere deeply pitted and eroded from the destruction of the more superficial and living part of the bone. The upper limit, at the line where the actual detachment has been effected, is highly jagged and tapering.

P. 603. A sequestrum of the femur from an amputation stump, measuring 18·5 cm. ($6\frac{1}{2}$ inches) in length.

At the sawn end it includes the whole of the bone; but beyond, only the deeper portions of the compact wall, the more superficial parts having been removed by erosion. The proximal end is the narrowest of all, the sequestrum ending in three long jagged spikes of different lengths.

P. 604. A long sequestrum detached from the femur after amputation.

It has characters like the foregoing except that the ring of bone at the sawn surface is at one spot incomplete.

P 605. A sequestrum 19 cm. ($7\frac{1}{2}$ inches) in length from the shaft of the femur, the necrosis having been due from infective inflammation following amputation. At its lower extremity it includes the entire thickness of the bone; elsewhere the superficial layers have been removed before the onset of suppuration, the further from the sawn surface, the more deeply so. The upper border along which the detachment has taken place is exceedingly jagged and irregular.

P. 606. Five large sequestra, exfoliated after amputations of thighs. They all include the whole, or nearly the whole, of the shaft at their sawn extremities, and all become smaller higher up. They all include less of the medullary tissue than of the walls of the shaft. There are deposits of new bone on the surfaces of some of them.

6. *Acute Osteomyelitis*
P. 607-624

Ulna

P. 607. A thick irregular case of new bone formed around a sequestrum, which comprises the greater part of the shaft of an ulna. The new bone formed upon the articular portions of the ulna which did not perish, and that formed around the sequestrum, have coalesced into one cylinder, which is complete with the exception of several round and irregular apertures in its posterior part. In both the articular surfaces of the ulna there are apertures communicating with cavities containing sequestra.

Tibia

P. 608. Part of a left tibia, of which portions of the wall have undergone necrosis. Almost the whole length of the shaft is covered with a thick deposit of new subperiosteal bone, which is perforated by cloacae in the situation of the sequestra.

P. 609. A tibia of which nearly the whole shaft suffered necrosis. The sequestrum comprises all the length of the shaft with the exception of about an inch at each extremity, and in some situations, includes the whole thickness of the wall. A considerable quantity of new bone is formed behind the sequestrum, but none anteriorly, except in one situation, a little below the tuberosity, where a thin scale is attached to the surface of the sequestrum. It is not improbable that this and some smaller portions near it were deposited on, and became adherent to, the surface of the bone before it had perished.

P. 610. A tibia of which a portion, about ten inches long, comprising part of the cancellous tissue of the head and nearly the whole thickness of the shaft, suffered necrosis. The sequestrum is completely separated, but is held in its place by the incomplete case of bone composed of the adjacent surviving parts of the shaft and the new bone formed on them. A small ulcerated opening leads into the knee-joint. On some parts of the sequestrum are portions of new bone, probably deposited before its death.

P. 611. Part of a tibia from which a portion of the anterior wall has separated after necrosis. The sequestrum lies loose in a cavity the front of which is formed of new bone and is perforated by several cloacae. The anterior part of the shaft is enlarged by depositions of new bone.

P. 612. Part of a tibia, of which a large portion, including its lower articular surface, has been nearly separated after necrosis. Previous to its death, the portion of bone was deeply and irregularly ulcerated, the inferior and external articular surfaces being destroyed and only a small portion remaining of the internal malleolus. There is little formation of any involucrum.

P. 613. Part of a tibia, in which necrosis has affected a portion of the cancellous tissue near its lower extremity. The dead bone lies loose in a large cavity within the shaft, which opens by several wide orifices through the walls, and through the articular surface into the ankle-joint. The whole of the articular surface is destroyed. The wall of the bone around the cavity is thin, light and porous; above it the walls are thickened and indurated, and appear as if there had been a large ulcer of the soft parts on the front of the leg.

P. 614. A sequestrum, about six inches long and, through two inches of its length, comprising the whole thickness of the shaft of a tibia. It was marked "Exfoliation from the tibia of a girl 8 years of age, which was supplied with new bone, and now is well and walks." (Old label.)

P. 615. A child's tibia almost the whole shaft of which has undergone necrosis as the result, probably, of acute osteomyelitis. The necrosis had involved the whole thickness of the shaft except the crest and a linear area of the posterior surface. A partial involucrum has formed around the sequestrum, and new bone has formed on the upper extremity of the shaft.

P. 616. The tibia of a child in which central necrosis has occurred at several spots within the shaft, as a result of acute osteomyelitis. The cavities in the bone contain several small sequestra and communicate with the exterior by small rounded apertures. The surface of the shaft is covered, almost entirely, with an irregular deposit of new periosteal bone.

P. 617. A tibia of which the whole of the shaft has undergone necrosis as the result of acute osteomyelitis. The necrosed bone has separated completely, and is surrounded on its external and posterior surfaces by a stout involucrum of new bone, which is perforated in several places by cloacae; over the inner aspect the involucrum is entirely wanting. At about the middle of the sequestrum is a thin layer of new bone, deposited before the occurrence of necrosis.

P. 618. Part of a tibia, of which a large portion of the shaft, including more than half its anterior wall, has undergone necrosis, probably as the result of acute osteomyelitis. The dead bone has separated completely and is retained within a cavity bounded by the remaining portions of the shaft and by a deposit of new bone. The cavity opens on the surface by one large, elongated aperture and by several smaller cloacal openings.

P. 619. A tibia of a young subject, almost the whole of the shaft of which has undergone necrosis, resulting from acute osteomyelitis. The dead bone has not completely separated, being still connected with the epiphyses and, in the lower part of the shaft, with the superficial layer of the cortex, which has escaped necrosis and is covered with a layer of new periosteal bone. The involucrum is very incomplete, especially over the upper part of the shaft. At the posterior and outer part of the upper extremity the inflammation has extended through the epiphyseal line, and the articular surface is partly eroded.

Drawing No. 96.

P. 620. A tibia, a large part of the shaft of which has undergone necrosis, probably as the result of acute osteomyelitis. New bone has been thrown out, forming an involucrum, which is widely open on the inner and posterior aspects. Near the upper end of the shaft the continuity of the bone is maintained only by a narrow strip of the involucrum about $\frac{1}{2}$ inch in width.

P. 621. Part of a tibia the lower end of which is riddled with cavities, some of which open on the surface of the shaft, some on the lower articular surface of the shaft, some on the lower articular surface. In one of the upper cavities is a small sequestrum and it is probable that sequestra have separated from the adjacent cavities. The lower extremity of the bone which is abnormally porous from rarefactive inflammation, is somewhat displaced outwards probably as the result of the porous condition of the bone. Drawing No. 77.

P. 622. A tibia of which a superficial portion of the shaft at its lower end has undergone necrosis. The sequestrum is enclosed in a cavity formed from new bone, which is widely open at its inner part, and new bone has been abundantly deposited on the surface of the shaft around.

P. 623. A tibia, part of the lower end of the shaft of which has undergone necrosis. A large quantity of new bone has been thrown out around the sequestrum and on the adjacent surface of the shaft, which in its upper part is superficially ulcerated.

Fibula

P. 624. The lower end of a fibula of which part has separated after necrosis. The sequestrum is loose in an incomplete cavity chiefly formed of new bone. The whole of the lower end of the fibula is enlarged.

7. *Necrosis*
P. 625-670

Skull

P. 625. The upper part of a skull, exhibiting superficial ulceration of the outer table of the frontal bone and part of the left parietal. The disease was probably connected with suppuration between the bone and the pericranium, the consequence of injury. The ulceration has, for the most part, proceeded uniformly through the superficial layers of the external table of the skull, leaving only a few small irregular portions, which retain their healthy texture and smooth surface and are elevated, like islands, on the surrounding rough and corroded bone. On the left frontal prominence, there are several small irregular apertures which involve the whole thickness of the skull. The bone thus ulcerated, died and was in progress of separation. At its lower part it is bounded by a deep ulcerated groove, which in many places has penetrated through the skull.

P. 626. Portion of a skull, in which a large piece of the frontal bone perished in consequence of a blow. The texture of the dead bone is not changed; but at the margins of the living bone the apertures for the blood-vessels are larger and more numerous than is natural, and in places an irregular but very shallow groove is formed.

P. 627. The upper part of a frontal bone, of which a considerable portion of the outer table is in process of exfoliation.

The dead portion, which has the mark of a blow on it, is nearly everywhere surrounded by a shallow groove of demarcation, the vascular foramina in the floor of which are abnormally large. Drawing No. 83, fig. 2.

P. 628. Part of a frontal bone, a small portion of the outer surface of which was in process of exfoliation. The dead bone is not altered in appearance, but is surrounded by a broad shallow groove and by large apertures for vessels in the adjacent living bone. The skull is fissured just below the dead portion.

The patient died some weeks after receiving a blow on the head. Soon after the injury a part of the scalp was reflected for the purpose of trephining; but no fracture being discovered, the trephine was not applied. The exposed portion of the skull perished; and while its exfoliation was proceeding the patient died. Drawing No. 83, fig. 1.

P. 629. Two large sequestra from the vault of a skull, involving the whole thickness of the bone. The bones are extensively pitted and worm-eaten in appearance. In the smaller fragment the outer surface is smooth, the disease being confined to the diploë and inner table. Drawing No. 87.

P. 630. A number of sequestra from skulls. Many of them show extensive superficial erosion of the inner or outer tables or both, as the result of acute suppuration. Some exhibit a worm-eaten appearance of the osseous tissue, with a formation of new bone or sclerosis of the pre-existing bone.

P. 631. A large portion of a parietal bone which was exfoliated after being ulcerated on both surfaces.

P. 632. A calvaria in which a portion of the external table of the left parietal bone has suffered necrosis and is in process of exfoliation. In most parts the necrosed bone is bounded by a groove of demarcation and the bone around is slightly thickened and porous. There is a trephine opening in the centre of the necrosed bone. The inner surface of the calvaria is normal.

P. 633. The posterior part of a skull from which a large portion of the right parietal bone has separated, probably in consequence of necrosis, leaving a perforation about 2 inches in diameter. The adjacent surfaces of the left parietal and occipital bones show irregular depressions, resulting probably from previous ulceration, and the whole of the affected surface and edges of the aperture are smoothly healed. Drawings Nos. 84, 85.

P. 634. A calvaria from which a portion of the left parietal bone has separated, probably after necrosis. The margins of the aperture are smoothly healed, and the surface of the bone around is slightly thickened and irregular on the surface.

P. 635. A calvaria, from which a portion of the whole thickness of the right half of the frontal bone has separated probably as the result of necrosis. The margin of the aperture is smoothly healed and the aperture closed with a fibrous membrane. The bone, for a considerable distance around the aperture, is thickened both externally and internally and about 1 inch above the upper margin of the aperture is a small irregular depression, possibly resulting from the healing of an ulcer. In the posterior parietal region the outer table is somewhat thickened and irregular.

P. 636. The posterior half of the skull of an Ass. There is an opening in the situation of the lambda, which appears to have been made with the trephine. The resulting aperture has been reduced in size by the growth of new bone from its margin.

This experiment may have been performed to show the slight osteogenic power of the flat bones of the skull.

Jaw

P. 637. A sequestrum consisting of the symphysis and adjacent parts of a lower jaw. The bone is irregularly excavated by ulceration.

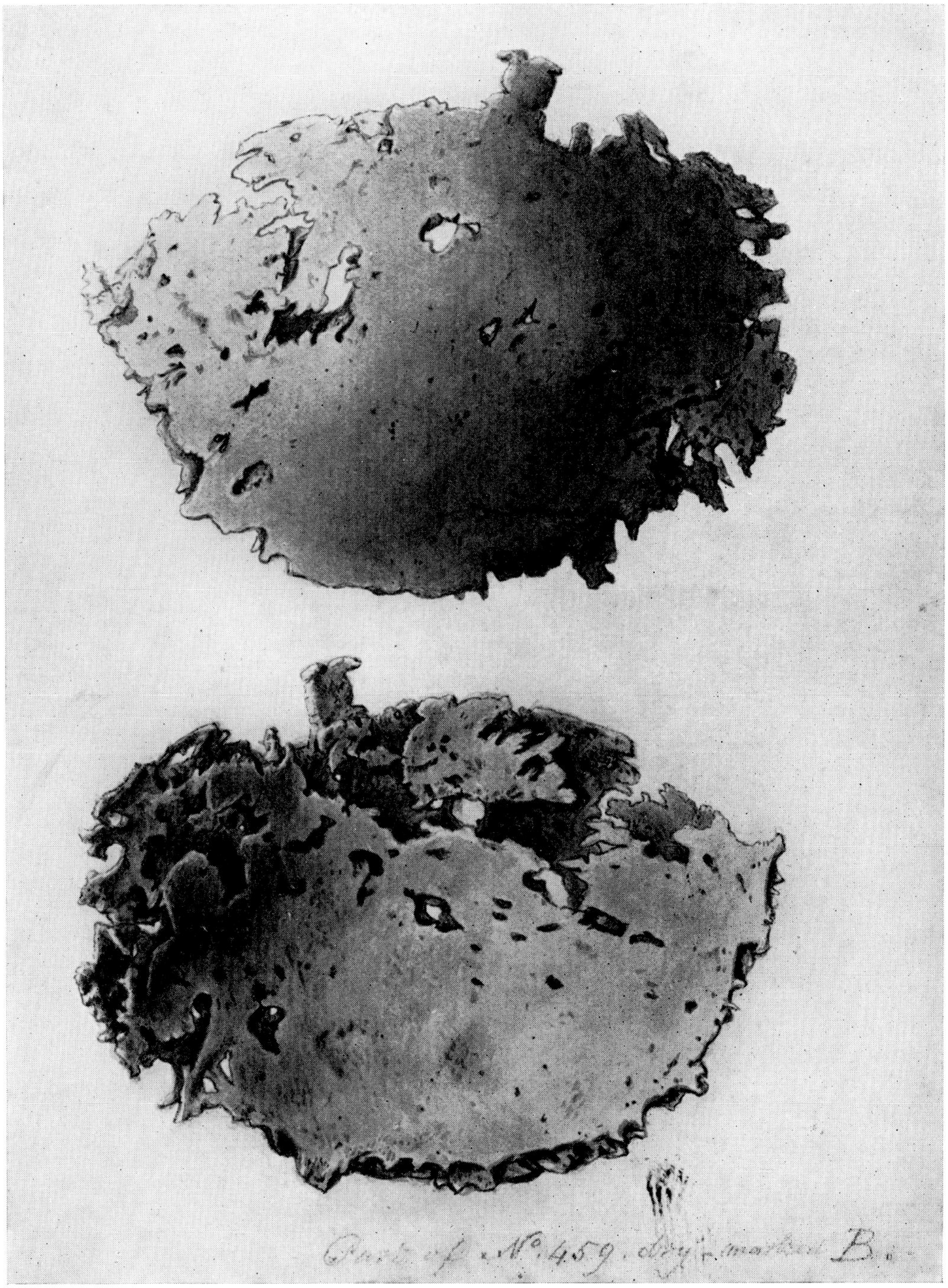

FIG. 6

"An external and an internal view of an exfoliation of part of the Os Frontis where the intermediate softer part of the Bone or Diploe is almost entirely destroyed by ulceration, separating the two tables of the skull from each other through almost the whole extent." Drawing No. 87 by William Bell. Specimen 629.

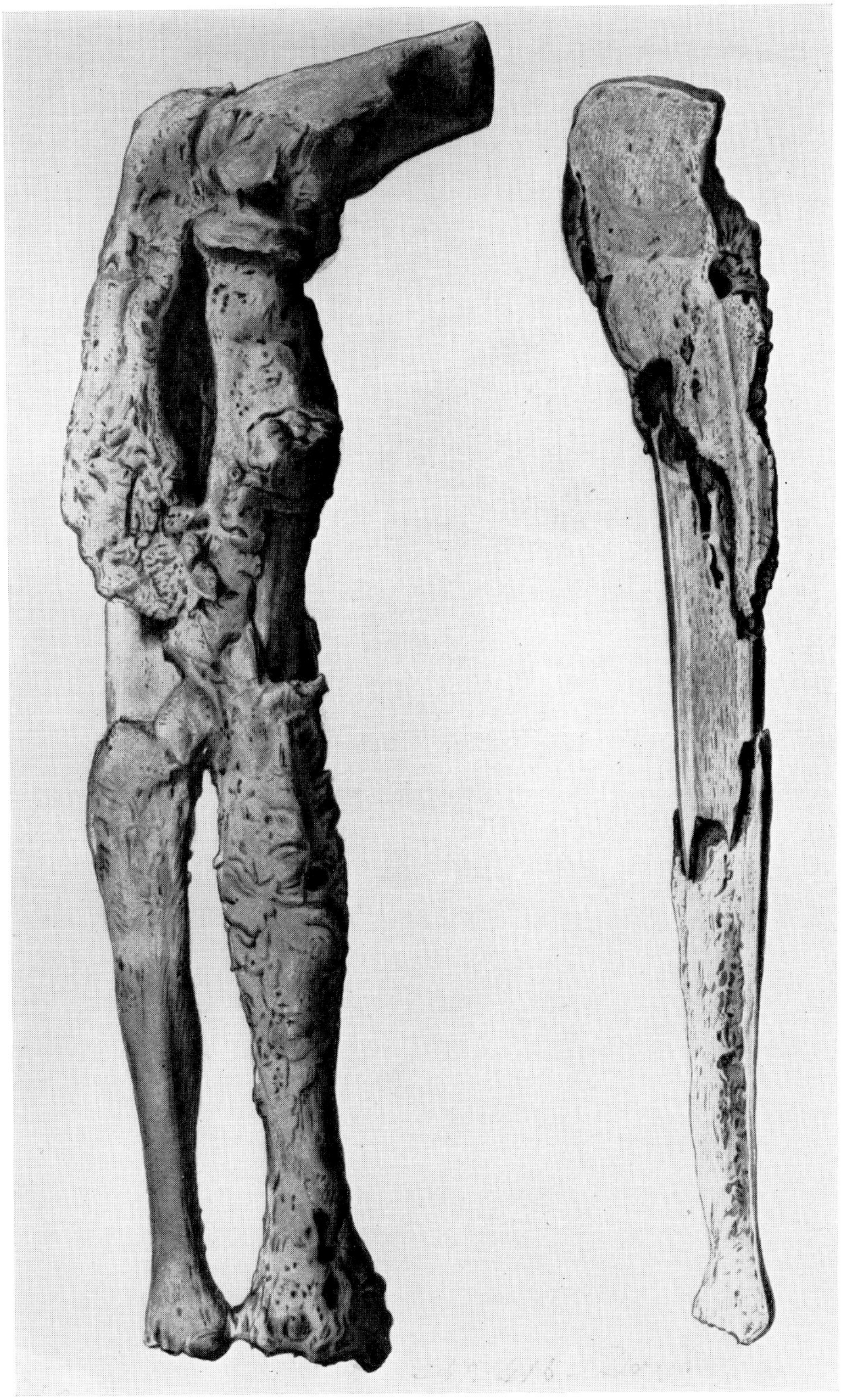

Fig. 7

(*a*) "The Ulna and Radius where both had become dead through their whole diameters in the middle between the two ends and by the time that the separation was completed, Nature had set up a bony continuity between the two extremes, as a substitute for the bone that was lost to the arm. The new bone formed on the two sides of the Radius and the Ulna which is nearest together and was so luxuriant as to form but one piece to the two, making a lateral anchylosis."
(*b*) "A view of the Ulna after being sawed down; looking upon its cut surface." Drawing No. 95 by William Bell. Specimen 640.

FIG. 8
"Death in a Whole Bone. Two views of a Tibia which had become wholly dead, from the epiphysis at the Knee to the joint at the Ankle; to supply the place of which Nature was encasing it round with living bone, beginning first at the two extremes and shooting towards the middle, but had not joined." Drawing No. 96 by William Bell. Similar but not identical with 649.

P. 638. Part of the body of a lower jaw which has suffered necrosis and separated as a sequestrum. The bone shows evidences of slight rarefactive inflammation, and there is a thin layer of new bone, chiefly on the posterior surface. The lower border is discoloured, probably from the application of the actual cautery.

Humerus

P. 639. The lower part of the shaft of a humerus of a young subject, from which a portion has undergone necrosis. The cavity which contained the sequestrum remains wide open. The rest of the shaft is altered in shape by superficial formations of new bone.

Radius and Ulna

P. 640. The bones of a forearm, with part of the humerus. Large portions of the shafts of both the radius and ulna have undergone necrosis. The necrosed portions have become enclosed in involucra, which are widely open superiorly, but inferiorly open on the surface only by small cloacal apertures, one of which communicates with the wrist joint through the radial articular surface. The two involucra are united by bone in the upper third of the bones. There is extensive osseous ankylosis of the elbow joint, in a semi-flexed position.
Drawing No. 95.

Ulna

P. 641. The bones of a right elbow-joint, showing a large cavity in the upper part of the olecranon probably resulting from caries and central necrosis. The wall of the cavity consists of porous bone showing little or no sign of healing, and near the upper end is a spongy sequestrum, still adherent to the bone and involving part of the sigmoid cavity of the ulna. Behind the sequestrum is a large aperture through which the carious cavity communicates with the elbow-joint. The upper extremity of the ulna is covered with osteophytes on its anterior surface and is ankylosed by bone to the humerus, in a flexed position. The osseous tissue of the trochlear surface of the humerus is rarefied. The radius is firmly ankylosed to the humerus, the lower extremity of which is thickened by a deposit of new bone. Presented by Sir William Blizard.

Femur

P. 642. A vertical section of the lower end of a femur of a young subject, in which there has been necrosis of a small portion of the cancellous tissue. In the middle of the shaft is a cavity, from which a wide canal leads obliquely downwards and forwards through the walls. The cavity is bounded by sclerosed cancellous tissue; and a few small portions of dead bone are adherent to its interior. The shaft around the cavity is enlarged by thickening of its walls, and plates of new bone are formed upon its exterior.

P. 643. Vertical sections of the middle of the shaft of a femur, exhibiting a small cavity in its anterior wall, from which a sequestrum has been removed. The anterior wall of this cavity is formed by new bone and opens externally. The adjacent parts of the shaft are enlarged and increased in density; and new bone has been produced on the surface.

P. 644. The upper part of a femur. There has been superficial ulceration of portions of the posterior and inner parts of the neck and of the shaft for about 4 inch below the trochanter minor. New bone has formed irregularly around the ulceration and in considerable quantity upon the anterior surface of the corresponding part of the shaft. The vascularity of the rest of the shaft was increased as evidenced by the presence of numerous minute foramina.

P. 645. The lower part of a femur, in which it is probable that there has been necrosis of a portion of the cancellous tissue. At the lower part of the shaft is a large cavity, which probably contained a portion of dead bone. This cavity opens widely through the posterior wall, and by a narrow aperture through the anterior wall; below, it appears to have opened through a long channel into the knee-joint. The whole shaft is very much enlarged. At the transverse section, about 8 inches above the condyles, the medullary tube is only $\frac{1}{4}$ inch in diameter, but the wall is, at the posterior part, nearly an inch thick. The whole substance of the walls is composed of dense cancellous and partly sclerosed bone; the exterior is in some parts smooth, but in others is rough and porous. There has been ulceration of portions of the articular surfaces and of the posterior part of the condyles.

Tibia

P. 646. A vertical section of the lower half of a tibia which has been decalcified, exhibiting necrosis of a large portion of the cancellous tissue involving the centre of its articular surface. In this section the lowest part alone of the dead bone remains. A bristle is passed through it; and it is further distinguished by the peculiar aspect of its lamellae, from which, without any apparent change in themselves, the endosteum has been removed. The dead bone is surrounded by a layer of soft vascular granulations formed in the adjacent cancellous tissue. A cavity, lined by similar granulations, extends for more than 3 inches up the interior of the shaft; within this cavity the remainder of the dead bone was enclosed. The wall of the tibia is thickened, and its external surface is irregular.

P. 647. A section of the shaft of the same tibia. The whole of the bone is greatly thickened and sclerosed, and a portion of its cancellous tissue, measuring 3 inches in length, has undergone necrosis. The cavity in which the sequestrum lies communicates with the surface of the exterior through a cloaca towards its lower end. Lower down in the shaft is a smaller sinus in the skin, marked by a bristle. The injection shows the walls of the cavities containing the sequestra to be highly vascular.

P. 648. Part of a tibia, on the outer and posterior aspects of which two portions of the walls have perished.

The upper, about 3 cm. ($1\frac{1}{5}$ inch) in length, is completely surrounded by a shallow groove of demarcation. The lower area of necrosis is very extensive; and its inferior limits are not yet similarly declared.

The surface of each sequestrum is quite smooth; and there is scarcely a trace of new bone having been formed on the shaft around.

P. 649. The shaft of the tibia of a young subject, in which necrosis has affected several isolated portions of the surface near its upper end. The adjacent parts of the bone are rough, partly as the result of ulceration, partly from deposits of new bone.

P. 650. Part of the shaft of a tibia from which large portions of the compact and cancellous bone have been removed after necrosis, either by operation or disease. The margins of the orifice in the bones are smoothly rounded, the whole bone is markedly porous from chronic inflammation, and the external surface exhibits a slight osseous deposit in places.

P. 651. Part of the shaft of a tibia, portions of the whole thickness of the inner wall of which have exfoliated. The whole of the external surface of the bone is eroded and the osseous tissue of the inner wall rarefied and ulcerated in its lower part.

P. 652. A sequestrum $5\frac{3}{4}$ inches in length from the front of the shaft of the tibia. The actual cautery and trephine have been employed to accelerate its separation.

P. 653. A transverse section from the middle of the shaft of a tibia, injected.
Portion of the bone is perfectly white, uninjected, and dead. The sequestrum is, for a considerable extent, separated from the living, by a zone of new formed, vascular granulation tissue.

P. 654. Two sequestra, exfoliated from the front of a tibia. The larger sequestrum consists of a portion of the outer layers of the shaft thickened, increased in density, and partly covered with new bone. In its centre is a circular aperture with smoothly rounded borders, immediately beneath which, and exactly adapted to a cavity in its under surface, the smaller sequestrum lies. It is probable that a portion of the inner wall of the tibia, comprised in the smaller sequestrum, first suffered necrosis and, exfoliating inflammation, gave rise to thickening and other changes of the bone around it, as well as to the ulceration of the bone immediately over it; the necrosis of the larger piece was possibly the result of its being exposed by ulceration of the soft parts over it.

Fibula

P. 655. A section of the lower end of a fibula, from which a portion of the malleolus was in process of exfoliation. The dead and living portions of bone are separated by a deep groove on their outer surface. The exterior of the fibula above the exfoliation is superficially ulcerated, and covered with new bone.

P. 656. The other section of the same fibula, macerated and dried.

P. 657. Part of a fibula including upwards of 4 inches of the whole thickness of the shaft, which, having suffered necrosis, was in process of separation. The shaft above the sequestrum is enlarged by the formation of new bone upon its surface.

P. 658. An irregular sequestrum about 3·8 cm. ($1\frac{1}{2}$ inches) in length from a fibula. Accompanying it there is a hemitubular piece of new bone which formed part of the sheath, and which subsequently underwent necrosis and was exfoliated with it.

P. 659. The lower part of a fibula, in which the whole thickness of the shaft for about $1\frac{1}{2}$ inches of its length is wanting, having undergone necrosis. The resulting gap in the bone has become partly repaired by the formation of columns of new bone uniting the fragments.

P. 660. The lower part of a fibula, a portion of the shaft of which has suffered necrosis, apparently after fracture. The dead bone is separated, but is held in its place by new bone formed around it.

P. 661. The lower part of a fibula, the lower extremity of which has undergone necrosis and is in process of separation.

P. 662. Part of a fibula. About the middle of the shaft a narrow groove passes irregularly around the wall, indicating that the bone below it has suffered necrosis. The other boundary of the dead portion is not obvious; but around the external malleolus there are several perforations as if produced by ulceration of the walls. Above the groove is a thin layer of new bone.

P. 663. A similar specimen, except that the separation of the sequestrum is not complete.

P. 664. The upper extremity of a fibula in which a total necrosis of the upper part of the bone has occurred, with the formation of a complete involucrum of new bone. The cavity communicates with the surface by several cloacae.

Os Calcis

P. 665. A left os calcis, the whole osseous tissue of which has an open porous texture due to rarefactive osteitis resulting probably from some chronic granulomatous disease. The bone is also riddled with sinuses, in some of which there are small sequestrae. There is no new bone formation except on the under surface and tuberosity, which is covered with osteophytes. The upper articular surface is partly destroyed by caries.

Metatarsal

P. 666. A metatarsal bone of some animal, the greater part of which has become expanded probably by a central, chronic inflammatory deposit. The wall of the shaft is thin, porous from rarefactive osteitis, and perforated in many places by small round apertures. A few irregular osteophytes are present, especially on the proximal end of the bone.

Phalanges

P. 667. Part of a great toe, in which several small portions of bone have been exfoliated from the distal extremity of the metatarsal bone. The bone surrounding the requestrum has been tinted a red colour.

P. 668. The extremity of a thumb, divided vertically. The distal phalanx, separated after necrosis, lies in a cavity surrounded by granulations, part of which separate it from the articular surface of the proximal phalanx.

"The joint was thus prevented from communicating with the dead bone, which might therefore have been removed without opening the joint or hurting the second bone." *MS. Catalogue.*

P. 669. One of the phalanges of a finger. The middle of the shaft is absent, owing to inflammation and necrosis. Attached to the proximal fragment is a small necrosed portion of the shaft. There is a considerable deposit of new spongy osseous tissue on the surface of both fragments.

Bone Fragments voided in Urine

P. 670. Several small portions of bone, marked "Bones voided by urine, after a blow on the lower part of the back, by Mrs. Wiltshire, recovered. Dr. Lewis attended her during the illness." They are chiefly thin scales of compact bone.

8. *Chronic Osteomyelitis in Leg Ulcers*
P. 671-691

P. 671. A left tibia. The inner surface of the lower half of the shaft is slightly flattened, probably owing to the separation of a sequestrum in this situation. The bone around this area and the adjacent part of the shaft is covered with a deposit of new bone. The affected area probably lay in the floor of an ulcer.

P. 672. Part of the shaft of a tibia. On the inner, subcutaneous surface there is an oval, elevated area of new bone, the uniform flatness and sharp edge of which show it to have been produced in the base of an ulcer of the overlying soft parts. The surface of the elevation is closed with a thin layer of compact tissue, an indication that the ulcer had cicatrised.

P. 673. Part of a tibia, from the anterior surface of which a thin oval portion of the wall was exfoliated, after being ulcerated and irregularly thickened. The borders of the shallow cavity left after the exfoliation are bounded by healthy new bone, beneath a part of which are two small sequestra, the separation of which is not completed.

P. 674. The portion of dead bone separated from the tibia last described.

P. 675. A tibia on which there is a circumscribed elevation of new bone formed in the base of an ulcer of the integuments. The new bone is becoming hard and condensed, indicating that the ulcer had healed. There is a general deposition of new bone on most of the rest of the shaft.

P. 676. A tibia and fibula, the shafts of which are enlarged and roughened from chronic osteoplastic periostitis.

On the inner, subcutaneous aspect of the tibia there is a well defined, slightly elevated oval area about 6·5 cm. ($2\frac{1}{2}$ inches) in vertical diameter, the surface of which is finely porous, and which has been produced in the base of an ulcer of the integuments. A short distance below it, there is a second smaller area of a similar kind.

P. 677. The lower two-thirds of a tibia. On the inner, subcutaneous aspect there is a long area of which the surface is formed by new bone, disposed in fine upstanding interconnected processes so as to produce an open network, the new tissue having formed in the granulations of an ulcer of the overlying integument.

New bone of the ordinary kind has been produced on the outer and posterior surfaces for a corresponding distance.

P. 678. A left tibia, the shaft of which in almost its whole length is covered by a thick deposit of new subperiosteal bone (osteoplastic periostitis). At several spots small areas of the surface of the shaft have undergone necrosis. The sequestrum lying above the internal malleolus is slightly discoloured and possibly lay in the base of an ulcer.

P. 679. Section of a tibia and of part of the surrounding tissues. The inner surface of the shaft, through a length of about six inches, is covered with new bone, in a layer which gradually increases in thickness from its margins to its central part, where it is rather more than half an inch thick. The greater part of the new bone appears harder and more compact than the original wall: they are intimately united; but a boundary-line between them may be discerned marking the unaltered outline of the wall. The new bone is also distinguished from the old by having a vertical grain, as if from vessels running through it perpendicularly to the axis of the shaft. The periosteum covering the new bone is slightly thickened; and the integuments over it appear to have recently healed after ulceration.

P. 680. The other half of the preceding specimen, macerated and dried. There is a deposit of new bone on about 6 inches of the inner surface of the shaft. The new bone is mostly smooth and compact, but at its upper part there is a well defined oval area about 3 inches in length in which the osseous tissue is irregular, although healed over, and which probably marks the situation of a healed ulcer. *Drawing No.* 76.

P. 681. Section of a tibia on which, in consequence of a large wound (the cicatrix of which is shown in the integuments), new bone has been abundantly formed over the whole or the greater part of the circumference of the shaft. A small part of the inner wall at the seat of the wound was removed. The outline of the original wall is but just discernible, the new bone formed on its exterior having completely coalesced with it. Opposite the wound the medullary cavity is also filled by new bone which has coalesced with the inner surface of the wall. The periosteum, a portion of which is reflected, is thickened, and appears to have been firmly adherent to the new bone.

P. 682. The lower part of a tibia. From the inner subcutaneous surface, an oval piece of the more superficial part of the compact wall is in process of exfoliation after necrosis. The dead part is undermined by a deep groove of ulceration. There had long been a deep ulcer of the superjacent integuments. The charring of the bone is due to the circumstances that when it was exposed the actual cautery was applied with the object of hastening the separation of the sequestrum. Drawing No. 88, fig. 2.

P. 683. Part of a right fibula, at the outer part of the lower third of which there is an extensive formation of subperiosteal new bone. The central part of the deposit is slightly concave and porous in texture. Immediately surrounding it is an abundant formation of osteophytes; and, higher up the shaft, a thin layer of compact new bone.

P. 684. Part of the left tibia of a young subject, the lower end of which is considerably enlarged, and bears on its inner wall a circumscribed oval elevation about $3\frac{1}{2}$ inches in length of soft and brittle new bone, with a slightly concave, finely porous surface and an abruptly rounded margin overhanging the adjacent part. New bone of compact texture has been abundantly formed on the surface of the shaft around. The circumscribed elevation of bone was, doubtless, formed beneath a chronic ulcer.

P. 685. A tibia showing changes very similar to those in the preceding specimen, except that the raised osseous deposit, corresponding to the base of the ulcer, is somewhat smoother, as if healing had commenced.

P. 686. A tibia, on the anterior wall of which there is a long oval superficial ulcerated area which probably formed the base of a chronic ulcer of the integument. The ulcerated surface is finely cancellous and porous. Around it new bone is irregularly formed upon the surface of the shaft.

P. 687. The lower half of a fibula nearly the whole of which is enlarged by a deposit of hard subperiosteal new bone. At the lower part of the outer surface there is a depressed area in which the superficial layers of the shaft have been destroyed by ulceration and rarefactive osteitis. The ulcerated area probably lay at the base of an ulcer in the integuments.

P. 688. The lower part of a right tibia. The shaft, except at the lower end, is greatly thickened by chronic inflammation. On the inner surface is an area 8 inches in length and slightly concave at its lower part, which is covered by a layer of porous new bone, and corresponded to the base of an ulcer in the skin. At the upper part of the specimen there is an irregular cavity, which contained a loose sequestrum, and communicates with the surface by several cloacae.

P. 689. The lower part of a tibia. A superficial part of the inner surface, about 4 inches in length, has undergone necrosis and is separated from the bone around by a narrow groove of demarcation. The surrounding bone is slightly rarefied, and the necrosed portion is partly covered with a thin layer of new bone. The tibia "had often been ulcerated; and by being laid bare for some extent, the exposed surface became dead, and the process of separation was just beginning". Drawing No. 88, fig. 1.

P. 690. A left tibia showing a deposit of new bone on the middle of the inner surface, an oval area of which, measuring $1\frac{3}{4}$ inches in length and including a superficial layer of the shaft, has undergone necrosis. A thin coating of new bone also covers the posterior surface of the shaft.

P. 691. A right tibia, on the inner surface of which is an oval depression about 2 inches in length with a rough eroded base, resulting probably from superficial exfoliation. The adjacent part of the shaft is covered with a thin layer of new bone.

9. *Osteoplastic Periostitis*
P. 692-705

P. 692. The lower part of a right tibia the surface of which is for the greater part covered with a thin layer of new subperiosteal bone.

P. 693. A femur of which the lower part of the shaft is enlarged by the superficial formation of new bone around it (osteoplastic periostitis). The formation appears to have taken place in distinct stages, nodules of finely cancellous bone having been deposited on the smoother and more compact layer surrounding the shaft. Higher up, the middle of the shaft is slightly enlarged by a thin layer of compact bone which has completely coalesced with the original wall. Drawings Nos. 24, 25, 26.

P. 694. A right femur, of which a large portion of the shaft is thinly covered with new bone resulting from chronic osteoplastic periostitis, most of which has become hard, smooth and compact. This is especially the case on the inner aspect of the femur, where the new bone is transversely grooved in the situation of vessels. Some parts of the new bone retain a longitudinally grooved and porous or finely cancellous texture.

P. 695. A right tibia from the same person as the femur last described. Nearly the whole shaft is enlarged as the result of osteoplastic periostitis. The new bone formation has occurred principally on the inner surface of the shaft, and is hard, smooth and compact, and is marked in places by shallow transverse grooves. At the upper part of the inner surface is a small deposit of porous bone.

P. 696. Part of a humerus, much thickened by a deposit of new subperiosteal bone, which has become hard and compact. Drawing No. 27, fig. 1.

P. 697. A left tibia enlarged by subperiosteal formation of new bone on nearly every part of its shaft. Most of the new bone has become hard and heavy; it has completely coalesced with the old bone and a great part of its surface has become compact and smooth. The largest accumulation of new bone is on the middle third of the inner surface where it forms an uneven swelling, the surface of which is very hard and perforated by few apertures for blood-vessels. A similar but thinner layer of compact bone covers almost the whole surface of the shaft.

P. 698. A fibula, of which nearly the whole shaft is enlarged from chronic osteoplastic periostitis. The new bone is hard, compact and heavy, and is inseparably united to the surface of the shaft. The chief deposit is on the inner surface, where the new bone is irregular, with thick, short, imbricated plates of various form; on its outer surface the bone is nearly smooth, but impressed superficially by transverse grooves for blood-vessels.

P. 699. A fibula, of which the middle half of the shaft has undergone irregular fusiform enlargement due to the formation of subperiosteal new bone, which at all parts has become hard and compact.

P. 700. A fibula, of which the whole length of the shaft is similarly but somewhat less, diseased.

P. 701. A left fibula, of which the whole length of the shaft is covered with an irregular formation of subperiosteal new bone the surface of which is hard and compact. At about the middle of its posterior and outer surface there are transverse grooves for blood-vessels.

P. 702. A fibula, a great portion of the shaft of which is greatly thickened as the result of chronic osteoplastic periostitis and covered with irregular plates of hard new bone.

P. 703. A left fibula, showing a uniform curve of its shaft. On various parts of the shaft there are isolated thin deposits of new periosteal bone. Some are porous, others have undergone sclerosis and are smooth and compact. At the lower end of the shaft there has been some superficial erosion from rarefactive inflammation.

P. 704. The humerus of an Ostrich, nearly the whole shaft of which is considerably enlarged and increased in weight. Its external surface is smooth and, in parts, finely perforated for the passage of vessels.

P. 705. Two of the carpal bones of a large bird, probably an Eagle, irregularly enlarged about the middle of their shafts.

10. "*Greasy Heels*"
P. 706-713

The following specimens of "greasy heels" were placed on loan to the Dick Veterinary School in Edinburgh in 1962 and, so far as can be ascertained, were destroyed in a fire there: P. 706, P. 707, P. 708, P. 710, P. 711, P. 713.

P. 706. The last phalanx of the foot of a Horse, at the posterior angles of which, in consequence of long-continued inflammation with "greasy heels", a large deposit of new bone has occurred, forming two coral-like processes extending backwards from the lower surface of the phalanx. One large mass has become detached from the specimen. Drawing No. 20.

P. 707. The phalanges of a Pony's foot, with deposits of new bone, more abundant around their articular surfaces. The distal interphalangeal joint is completely ankylosed.

P. 708. The phalanges of the foot of a Horse, with abundant deposits of new bone. The second and third phalanges are firmly ankylosed by bone and at the side of the second phalanx is a deep cavity, marking the site of an abscess, from which a smooth, shallow groove runs downwards on the surface of the bone, resulting probably from a suppurating track.

P. 709. A longitudinal section of the first two phalanges of a Horse's foot united by bone formed around their corresponding articular extremities. The new bone is in large irregular masses, variously connected, which have a coarsely cancellous texture. The articulation between the phalanges has been destroyed and is undergoing ankylosis.

P. 710. The upper part of the right metatarsal bone of a Horse, surrounded by nodulated hard new bone at its articular extremity, and firmly united with the bones of the tarsus. The chief bond of union appears to have been formed by the coalition of the growths of new bone extending from the several bones over the margins of the joint.

P. 711. The lower end of the left radius, with the carpal and metacarpal bones of a Horse. The carpus is surrounded by large, irregular masses of new bone which has formed on the surface of the carpal and contiguous bones. Three of the carpal bones are firmly ankylosed to the radius, and the remainder of the metacarpal bone. The articular surfaces of the joints are healthy and the ankylosis has resulted from ossification of the soft structures around the joint.

P. 712. Several phalanges, probably of a Bear, thickened and covered with irregular formations of hard, nodulated bone. In one, the distal articular surface of the phalanx is eroded and the osseous substance has suffered some degree of rarefactive inflammation.

P. 713. The left metacarpal bone of an Ox, with the first phalanges and sesamoid bones. They are all firmly united by bone, formed around and between their adjacent articular surfaces.

11. *Syphilis*
P. 714-746

Skull

P. 714. The anterior half of a skull. On the right side of the frontal bone, in which the frontal suture is persistent, there is a somewhat circular area exhibiting changes which may be attributed to syphilitic disease.

Over the centre of the area the surface is raised and tuberculated from the coalescence of smaller eminences of new formed bone. At the margin, where the disease is extending, the inflammatory process had led as yet only to rarefaction of the outer table, the new bone elsewhere having been subsequently reproduced on the rarefied tissue. There is a much smaller lesion of the same kind on the left side of the frontal bone.

Corresponding with the sites of external disease, there are two areas of the inner table which are thinly invested with sclerosing osseous tissue.

P. 715. A skull, in the tabular portion of the frontal, and in the parietals of which there are a series of irregular, carious pits, attributable to syphilitic disease.

The ulcers have no definite form though the earliest appear as small circular apertures arising at the sites of vascular channels; by the local multiplication and coalescence of such, the larger lesions result. In many, the outer table is much undermined; and in a few spots the entire thickness of the calvaria is perforated. The surface of the frontal bone between the ulcers is slightly raised so as to be lowly tuberculated.

Of the bones of the face, portions of the nasal process of the right superior maxilla, and much of the palatine processes of the maxillae and palate bones, etc., have been destroyed by caries, combined probably with necrosis.

P. 715A. A calvaria in which, from syphilitic disease, nearly the whole outer table is superficially ulcerated. The ulceration presents an "annular" form; it appears to have commenced in isolated spots, and to have spread either in rings enclosing portions of healthy or ulcerated bone, or in circles; but in most places the ulcers thus formed have coalesced and become diffused, and have extended very irregularly in depth, in a few places small perforations having occurred. The greater part of the posterior half of the right parietal bone has been removed, probably after necrosis. In the lower left parietal, frontal and occipital regions the surface of the bone is irregular as the result of the healing of the carious surface.

P. 715B. A calvaria showing changes resulting from syphilis. On the right side a large irregular aperture, measuring 4 inches in its longer diameter, has resulted from necrosis. The surrounding bone is carious, that immediately around the aperture being ulcerated and bounded externally by very porous bone with caries of the diploë. At the vertex is a small ulcerated depression with smooth shelving and healed edges. On the left side of the calvaria is a large carious surface, almost symmetrical with that on the right side, which is perforated in several places and shows little signs of healing except at its lower part, where the ulcerated grooves have become coated with an irregular layer of compact bone. There is an extensive layer of porous new bone on the inner surface of the cranium.

P. 716. A skull showing changes resulting from syphilis. At its posterior part is a large area of caries, with numerous small perforations, in which no formation of new bone has occurred. In the frontal and upper nasal region the lesion is seen in a later stage: the greater part of the surface shows carious ulcers, but the intermediate bone has proliferated and extended into the depressions, narrowing them in some places, obliterating them in others, leaving typical bone-scars as in the bone immediately in front of the sagittal suture.

P. 717. A calvaria showing changes resulting from syphilis, affecting mainly the frontal and posterior parietal regions. At the bregma and posterio-inferior part of the right parietal bone there are irregular worm-eaten depressions resulting from syphilitic caries, the edges of which are sharply cut. At the lambda a similar carious area shows the base partly healed over with new bone, and above this is a large area in the parietal bones in which the ulcerated parts have been narrowed to mere grooves by the outgrowth of new bone from their margins. In the frontal region the process of healing is almost complete, producing a somewhat nodular grooved surface in which the new bone is still abnormally porous. On the inner surface of the calvaria there are three small depressions in the frontal, which were possibly caused by perforations; and there is a deposit of hard bone on either side of the middle line.

P. 718. A calvaria in which there is a circular ulcerated aperture, $\frac{2}{3}$ inch in diameter, in the middle of the left parietal bone; its margins are oblique and smooth. Anterior to it there is a small circular healed ulcer in the external table. On the opposite parietal bone is a circular irregularly penetrating carious area, $\frac{1}{4}$ inch in diameter. The greater part of the external surface of the calvaria shows signs of increased vascularity and there are slight deposits of new bone on the inner surface in connection with the perforations.

P. 719. A calvaria showing the results of syphilitic disease. In the frontal region there is an irregularly rounded, carious area, $1\frac{1}{2}$ inches in diameter. The central part is deeply ulcerated and perforated in several places. The periphery of the area is marked out by a sinuous groove, probably indicating that the carious bone has suffered necrosis. Almost the whole of the remaining surface of the frontal bone is rendered uneven by shallow depressions, indicating the position of carious areas which have become healed over with compact bone. Several depressions of a similar kind are present in the left parietal bone.

P. 720. A calvaria showing the results of syphilitic disease. In the left frontal region is a large irregular aperture measuring $2\frac{1}{2}$ inches in width, which has resulted from necrosis: the margins of the aperture have a worm-eaten appearance due to caries. In the right posterior frontal region is a smaller ulcerated and perforated area, the edges of which are shelving, smooth and partly healed over with compact bone. In the left lower parietal region is a small shallow depression in the position of a healed ulcer, the floor of which is marked by fine fissures. In the lower frontal region are one or two smooth depressions, marking the sites of previous ulcers.

P. 721. A calvaria showing changes resulting probably from syphilis and suppuration. At the vertex is a large ulcerated area with two perforations, due probably to necrosis. The bone surrounding the ulcer is porous as the result of caries and the adjoining surface of the left parietal bone is slightly irregular, thickened and marked by small grooves, indicating probably a healed carious area. The main part of the base of the ulcer consists of the exposed diploë, the external table having been destroyed over a continuous area, probably by suppuration; a similar destruction of the inner table has also occurred.

P. 722. Part of the base of a skull, in which nearly the whole of the hard palate and alveolar process of the upper jaw, together with portions of the septum nasi and of the nasal bones have been destroyed by syphilitic ulceration. The under surface of the body of the sphenoid bone is also ulcerated and the floor of the sphenoidal sinus perforated by an aperture, $\frac{1}{2}$ inch in width.

P. 723. Part of the calvaria of a young subject, which has apparently separated after necrosis, probably syphilitic. The bone is very thin and is covered, both internally and externally, by a thin deposit of new bone. The sutures are ossified. Drawing No. 86.

Long Bones

P. 724. A right ulna, with a left radius, apparently from the same patient, exhibiting changes due, doubtless, to syphilitic disease.

Much of the shaft of each bone is irregularly enlarged from chronic inflammation, and in these thickened regions there are many deep carious pits. In some of the latter the surface is covered with delicately cancellous, reparative bone, formed in the investing granulation tissue, the edges of the cavities being, moreover, in process of contraction from the outgrowth of new osseous tissue; in others, the original substance of the shaft, rarefied from caries, is exposed at the bottom.

From the larger cavities, sequestra have probably been detached.

P. 725. A right femur, of which the upper half of the shaft is enlarged by the presence of a subperiosteal node occupying chiefly its inner aspect. There is a similar but slighter enlargement of the lower third of the shaft. The surfaces of the thickened parts are smooth, compact, and very finely perforated.

P. 726. The lower end of a femur, of which the shaft is enlarged, chiefly by the superaddition of new bone. A transverse section shows that the new bone has, for the most part, intimately coalesced with the circumference of the shaft, but may be clearly traced all around it. The medullary cavity of the shaft is in process of being filled with new bone.

P. 727. The lower half of a femur, considerably enlarged. The enlargement appears to be chiefly due to the external formation of a thick irregular layer of bone, which has completely coalesced with the original shaft, and has a finely perforated surface marked behind by broad transverse vascular impressions.

P. 728. A section of a femur. The medullary canal at the junction of its lower two-fourths is filled with a mass of new spongy bone, in the centre of which is an irregular cavity. Above this, for about 3 inches, the canal is filled with dense spongy bone, which in places has become sclerosed. Externally the shaft is coated with hard new bone at the level of the disease in the medulla, and the shaft has become curved in this situation. It is possible that the lower part of the medullary canal was the seat of a gumma.

P. 729. Another section of the same femur.

P. 730. A femur, the lower third of which is enlarged as the result of osteoplastic periostitis. There is a large accumulation of porous new bone at the upper part of the popliteal surface; in many places the surface of the deposit has become hard and compact, and it is continued into a thin compact layer of new bone covering almost the whole surface of the shaft. On the outer tuberosity of the femur there is a shallow digital depression, the floor of which is smoothly healed, and probably marks the site of a healed gumma.

P. 731. A left femur the greater part of the shaft of which is covered with a layer of new subperiosteal bone.

P. 732. The opposite femur of the same person, similarly diseased. The new bone has an almost exactly symmetrical arrangement on the two shafts.

P. 733. A tibia from the same patient. Considerable parts of the surface of the shaft are covered with a thin layer of new bone (osteoplastic periostitis). On the posterior surface the new bone has a worm-eaten appearance in places due to caries, probably syphilitic.

P. 734. A right tibia. Almost the whole surface of the bone is covered with a layer of new osseous tissue. At the lower end of the shaft the deposit is more abundant and is excavated by several depressions, some of which extend deeply into the substance of the bone: one of the cavities contains a small sequestrum. The disease is stated to be syphilitic.

P. 735. Half of a tibia of which the lower third is enlarged from chronic inflammation.

On the inner, subcutaneous, aspect there is a deep, highly irregular, and undermined carious surface, which probably corresponded with a syphilitic ulcer of the integuments; the lesion is in process of healing, being covered with delicate upstanding processes formed in the investing granulation tissue, or closed in with a thin compact layer of new bone.

The caries was probably associated with some necrosis.

P. 736. A right tibia, of which from the inner, subcutaneous wall, a portion 12·5 cm. (5 inches) long has been completely detached after necrosis. In the centre of the sequestrum, there is an oval carious depression indicating that the bone was the seat of disease previous to its death.

The shaft for a wide extent is covered with new bone, and presents cavities here and there, in one of which there lies a small sequestrum.

P. 737. An oval scale of bone about 10 cm. (4 inches) in its chief diameter, which has separated after necrosis, from the front of the tibia.

On the exterior is shown the slightly depressed surface of an ulcer, in process of healing, and which corresponded with an ulcer in the integuments. The bone beyond the ulcer is coarsely grooved in the longitudinal direction, and the general convexity of the exterior proves that the sequestrum comprises portion of the shaft previously thickened from chronic syphilitic inflammation.

The necrosis was probably due to a secondary infective inflammation arising in connection with the ulceration of the integuments.

P. 738. A right tibia, nearly the whole surface of which is covered with a thin irregular formation of new bone which in places culminates in more pronounced elevations or nodes: the disease is presumably from a case of acquired syphilis.

P. 739. A left tibia, the greater part of the shaft of which exhibits changes due to chronic osteoplastic periostitis.

The shaft is, in places, thinly invested with new tissue which is, where thinnest, longitudinally grooved by the periosteal vessels. On the inner and outer aspects of the shaft the new bone is accumulated so as to form two lowly raised elevations, or nodes; that on the inner, subcutaneous surface being the more prominent.

The disease is probably syphilitic. The right tibia was similarly and symmetrically diseased.

P. 740. A right tibia, of which the middle half of the shaft shows a fusiform enlargement resulting from the formation of subperiosteal new bone. In most places the surface of the new bone is smooth, hard and compact. The disease is probably syphilitic.

P. 741. The outer half of a left tibia, showing a syphilitic node in section.

The bone composing it has become as dense as the subjacent compact wall of the shaft, with which it is inseparably continuous, and which it surrounds, though least thickly posteriorly.

P. 742. The other half of the same tibia.

P. 743. Part of a tibia in which a deposition of new bone covers nearly the whole of the shaft.

P. 744. A right tibia and fibula the shafts of which are enlarged by a deposit of new subperiosteal bone, involving in the case of the tibia, the greater part of the shaft, and in the fibula the upper third. Most of the new deposit has become hard and compact on the surface. The disease is probably syphilitic.

P. 745. A section of a tibia, the shaft of which is much thickened by chronic osteoplastic osteitis, probably resulting from syphilis. The upper half of the medullary canal is filled with spongy new bone in which are numerous small excavations caused probably by necrosis and communicating with the surface of the bone by several narrow cloacae. In the lower third of the canal is an incompletely separated sequestrum, lying just within a cloacal aperture half an inch in diameter. The surface of the shaft is covered with a thick layer of new bone, mostly compact, but in the upper part more porous.

P. 746. A fibula, the greater part of which is covered with a deposit of hard subperiosteal bone. Near the centre of the shaft is a fusiform swelling with a deep excavation in its central part, which in the recent state was probably occupied by gummatous tissue. Most of the osseous deposit on the shaft has healed smoothly over. From the same individual as P. 738.

12. *Rickets*

P. 747-762

Skull

P. 747. The frontal bone of an infant which was affected with rickets. The greater part of the bone is thickened, and covered with a light, finely porous and friable osseous tissue, the external surface of which resembles the surface of woollen cloth. The change of structure affects exactly symmetrical portions of the two lateral halves of the frontal bone, and appears to have made progress from its circumference towards its original centres of ossification; for these centres, together with a space of irregular shape around them, are free from the disease. On the inner surface of the bone the same change of structure, though in a less advanced stage, is more generally diffused.

P. 748. The upper part of the skull of a child, about 4 years old. The bones are much thickened throughout and converted into a finely porous osseous substance which has replaced both tables so as to give the exterior a porous in place of a smooth surface.

Long Bones

P. 749. The humeri, from an adult, of which, as a result of past rickets, the upper part of the shafts are slightly curved outwards. The tissue in each is firm and normal, and smooth as the shaft of a healthy bone.

P. 750. The left humerus of an adult, the whole shaft of which in consequence of rickets, is considerably curved outwards. Its tissue appears healthy, and the parts for the attachment of muscles are well developed.

P. 751. The femur, from a person of nearly adult age, the upper epiphyses being completely, the lower nearly united. It is similarly and symmetrically altered; and exhibits the effects of rickets partially recovered from. The upper and middle thirds of the shaft is curved forwards, and laterally compressed; and the linea aspera at this curved part projects far backwards, its posterior margin forming the chord of the arc which is formed by the anterior wall of the femur. All this part of the bone appears of healthy texture, but is hard and heavy. In some areas, the surface of the lower end of the shaft is slightly porous owing to persistence of abnormal bone produced whilst the disease was active.

P. 752. The other femur from the same person, showing similar and symmetrical changes.

P. 753. An adult femur, of which the shaft is more than usually curved forwards; its tissue, though rather light, appears healthy; and its neck has its natural obliquity.

P. 754. An adult femur, the shaft of which, in consequence of rickets, is a little more than usually curved forwards. The axis of the neck forms a right angle with that of the shaft, and the head is below the summit of the great trochanter; but the substance of the bone is healthy.

P. 755. An adult femur, similarly diseased. The linea aspera is prominent, and the inner border of the shaft is sharper than is natural. The internal condyle is placed more than usually backwards, and the external one is proportionally prominent, so that the lower part of the shaft looks as if it had been twisted. The osseous tissue appears healthy.

P. 756. A right femur, exhibiting almost exactly similar changes. The bone is small and slender in all its parts, especially at its articular portions; but its tissue appears healthy.

P. 757. The left femur of the same person, similarly and symmetrically changed.

P. 758. The right tibia and fibula of the same person. Both of them are curved forwards and inwards, and their shafts are much flattened, that of the fibula being, at one part, an inch wide and about $\frac{1}{4}$ inch in thickness. The tissue of both bones appears healthy; and their extremities are natural in form and size.

P. 759. The left tibia and fibula of the same person, similarly and symmetrically changed.

P. 760. An adult right tibia, the shaft of which is curved inwards in consequence of rickets.

P. 761. A tibia, of which the shaft, from the same disease, is curved forwards and inwards, and remarkably flattened, its inner and outer surfaces being in nearly parallel planes. The shaft is increased in density.

P. 762. A tibia and fibula, the shafts of which, in consequence of rickets, are remarkably curved inwards and flattened. At the most prominent part of its curve, which is just above the junction of its middle and lower thirds, the shaft of the fibula is $1\frac{1}{4}$ inches in width, but not more than two lines in thickness. Just below this part an irregular process of bone has grown from the outer border of the tibia, and is received in a corresponding cavity on the inner border of the fibula, so as to form a kind of false joint between them. There is a similar smaller growth on the inner margin of the tibia. In other respects the bones are healthy.

13. *Rickets-like Disease in Captive Animals*
P. 763-767

P. 763. A humerus of a rickety Lion. In the upper part of the section the compact wall of the shaft is in process of rarefaction, and in the lower part the rarefied bone is in addition invested with a layer of finely porous bone. There is some irregularity in the advancing end of the shaft at the lower intermediary cartilage.

P. 764. A humerus and part of the scapula of the same rickety Lion. Their whole texture is light, dry and friable; they have scarcely any compact tissue; and what there is, is very thin, porous, and brittle. Part of the scapula also has been broken or bent and is fixed in its unnatural position.

P. 765. Part of the femur of the same Lion, similarly diseased. The shaft has been fractured at the junction of its upper and middle thirds; and no union has been effected, although some very light and spongy new bone has been formed about the ends of the fragments. In the lower third of the femur is a green-stick fracture.

P. 766. The ulna of a Monkey, divided vertically. The whole bone is extremely porous, and a large part of the medullary canal is filled with new-formed spongy bone. The upper part of the bone is covered with a subperiosteal deposit.

P. 767. Numerous fragments of bones of the same Monkey. The osseous tissue is extremely light and porous and the bones are thickened, mostly by endosteal new bone production but in one or two instances by a subperiosteal deposit. The cranial bones are thick and consist of finely porous bone without distinction of diploë and tables.

14. *Hypertrophic Pulmonary Osteoarthropathy in Animals*
(*Symmetrical Osteoplastic Periostitis*)
P. 768-798

Note: The true nature of some of these specimens is uncertain. See also in Section II (5) Osteoarthritis (p. 104).

P. 768. The left tibia of a Lion. The surface is for a large extent invested with new periosteal bone.

Where thinnest the new formation is longitudinally grooved by the vessels lying in the surface of the original bone. In the further progress of the disease, as seen in other situations, the bone takes the form of nodules which appear upon the first formation, and then by lateral extension come to meet so as to produce a curiously patterned surface not unlike that of the skin in ichthyosis. The whole of the superadded bone is less white in colour than the original shaft, in consequence of its fine, vascular, porosity, secondary sclerosis not having as yet occurred.

P. 769. The right tibia of the same Lion, similarly diseased. The osseous formations, though very irregular, are arranged in precise symmetry with those on the left tibia.

P. 770. A section of the right femur of the same Lion after being decalcified by acid. Nearly its whole surface is covered by a nodulated osseous formation. It is closely connected with the surface of the old bone, although the line of boundary between them is very evident.

> "From this preparation of one of the bones of the Lion, which had been injected and steeped in acid, it appears that this newly added bone was vascular, and had only lost its earth by this steeping, its form still remaining: it therefore must be as much an animal substance as the old bone."
>
> *Works*, vol. i, p. 507.

P. 771. The other section of the same femur. Drawing No. 14.

P. 772. Sections of the left femur of the same Lion. Like the right femur, and with an exactly similar arrangement, a large portion of the surface is covered with new bone, the structure of which is nodulated, hard, and minutely cancellous. In the sections the boundary between the new and the old bone is quite distinct.

P. 773. The pelvis of the same Lion. Large portions of the outer surface of each innominate bone are covered with new bone, like that formed on the femora and other bones. The new bone is deposited with exact symmetry on corresponding parts of each lateral half. There is a similarly symmetrical formation on the inner surface of each ischium, near the obturator foramen.

Drawing No. 16.

P. 774. A patella from the same Lion, with slight deposits of new bone along its lower margin and upper border.

P. 775. The right os calcis from the same Lion, with new bone (like that already described) formed abundantly on nearly all its surface.

P. 776. The left os calcis, similarly and exactly symmetrically diseased.

P. 777. One of the tarsal bones of the same Lion, with some small nodulated formations of new bone on a portion of its surface.

P. 778. Part of the left scapula of an old and very large Lion. The acromion process is covered with thick, irregular, porous and channelled new bone, which overhangs its margins and is intimately united to its surfaces. The other parts are healthy.

P. 779. The left ulna of the same Lion. It is enlarged, and its surface is rendered uneven by formations of new bone, which have completely coalesced with the original wall and have become very hard, heavy, compact, and smooth on their surfaces. Nearly the whole shaft is thus covered; and part of the new bone projects in sharp ridges.

P. 780. The right ulna of the same Lion, similarly and symmetrically diseased.

P. 781. The right femur of the same Lion similarly diseased, but to a greater extent. The new bone is more abundant; like that on the ulnae, it is very hard, compact, and heavy; on the anterior part of the shaft its external surface is nearly smooth; on the posterior part it has an irregularly laminated and knotted arrangement, and portions of it project in sharp uneven ridges.

P. 782. The right tibia and fibula of the same Lion similarly diseased, and exhibiting a similar exact symmetry in the arrangement and structure of the new periosteal bone.

P. 783. The left femur of the same Lion similarly and symmetrically diseased.

P. 784. The left tibia and fibula of the same Lion similarly diseased. The interosseous ligament is ossified in nearly its whole length; and at the lower part the masses of osseous substance formed on each of the bones have coalesced.

Drawing No. 52.

P. 785. Two metacarpal bones of a Lion, probably the same as that from which the preceding bones were taken. They are united by new periosteal bone formed on their dorsal surfaces and borders. Drawing No. 51, fig. 2.

P. 786. Two of the tarsal bones probably from the same Lion, showing large masses of new bone projecting from their surfaces.

P. 787. A caudal vertebra from the same Lion, showing masses of new bone projecting from its surface.

P. 788. The right scapula and humerus of a Lion. Large masses of bone, of light and porous texture, have been formed about the notch of the scapula and the lesser tuberosity of the humerus. Drawing No. 67, fig. 2.

P. 789. The left tibia and fibula of a young Lion. They are united by new bone formed on the adjacent borders and surfaces of their lower halves. The rest of their shafts and their articular surfaces are healthy.

P. 790. The pelvis of an old Lion which had lived long in England showing changes due to chronic osteoplastic periostitis. Formations of grey and finely cancellous bone spring from several parts of its outer surface. They are irregular in form and thickness, but are placed with an almost exact symmetry on the two halves of the pelvis.

P. 791. The last two lumbar vertebrae and part of the sacrum of the same Lion, showing similar formations resulting from chronic osteoplastic periostitis.

P. 792. The right humerus of an old Lion which had lived long in England. There is an abundant deposit of periosteal new bone round the articular surface of the elbow-joint, which shows evidence of advanced osteoarthritis with eburnation.

P. 793. The right radius and ulna of the same Lion, showing similar changes.

P. 794. The left humerus of the same Lion, similarly and symmetrically diseased.

P. 795. The left ulna of the same Lion similarly diseased.

P. 796. Two dorsal vertebrae of a Lion. The vertebral bodies are atrophic and ankylosed by ossification of the anterior common ligament. Stout nodulated ridges of bone project from the margins of the bodies and the articular processes. The space for the intervertebral disc is much enlarged.

P. 797. The second to the sixth cervical vertebrae of a feline animal. From the lower margins of the laminae of the second and third vertebrae two plates of bone, evidently developed in the soft structures connected with the bones, are directed caudally. They overlap the succeeding laminae, which are covered with new bone. Drawing No. 66, fig. 2.

P. 798. The right scapula and humerus of probably the same animal. The borders of their articular surfaces, and other parts adjacent to the joint, are covered by thick irregular masses of light and finely cancellous new bone. The masses attached to the scapula and the humerus had, in one situation, coalesced outside the joint so as to destroy its mobility. The articular surfaces and the rest of the bones are healthy; but the head of the humerus has not its usual bearing upon the glenoid cavity. Drawing No. 67, fig. 1.

15. *Paget's Disease*
P. 799-803

Paget's Disease—Skull

P. 799. Portions of a calvaria uniformly thickened and of a finely cancellous structure throughout. The discoloration of the bone, and the partial loss of its surface suggest that it has been exhumed.

P. 800. The greater part of a skull, the bones of which are greatly thickened, due evidently to osteitis deformans. A large portion of the posterior and upper part of the skull has been broken away; and the edges of the bone thus exposed have a thickness varying from $\frac{3}{4}$ to $1\frac{1}{4}$ inches. The whole of the parietal and a great part of the frontal bones are of the same thickness: the squamous portions of the temporal bones are $\frac{1}{2}$ inch thick; but below them the thickness gradually decreases towards the middle of the base of the skull, where the bones are much less altered in size. The superior maxillary, malar, and sphenoid bones are thickened, but less than the bones of the cranium. The vomer, except its base, and the hinder part of the hard palate are not thickened. The tissue of the upper part of the skull is converted into a nearly uniform, finely porous cancellous bone. The external surface of the thickened bones is abnormally porous. In the bones of the face the same porous change is less distinctly marked. The cranium is flattened from above down and the transverse and antero-posterior diameters correspondingly increased.

The skull was found in digging a grave in the churchyard of Stepney, and was given to Mr Hunter by Mr Patton, surgeon, of Ratcliffe Cross.

(A large fragment from the occipital region and numerous smaller fragments from different parts of the skull, are preserved in the box in which the specimen is mounted.)

P. 801. A calvaria the thickness of which is greatly increased except in the left parietal region. The distinction between diploë and inner and outer tables is lost and the osseous tissue is throughout its thickness uniformly and finely porous in texture. The vascular grooves on the internal surface are deeper than normal. The sutures are entirely obliterated.

P. 802. A parietal bone, an inch in thickness. It was found in the Hunterian Collection of fossils, marked in a foreign handwriting, "Antropolithus, s. Craniū humanū, locus incertus." *Osteology Catalogue*, 1831, No. 119.

P. 803. A calvaria showing a uniform thickening. The distinction between diploë and outer and inner tables is lost and the whole thickness of the bone consists of dense, finely porous osseous tissue. The vascular grooves of the inner surface are deeper than normal. The sutures are obliterated.

16. *Myositis Ossificans*

P. 804-806

Generalised

P. 804. The skeleton of a man aged 39, who suffered from myositis ossificans. In the skull there are small bony outgrowths or osteomata in the following situations: the left supraorbital ridge, the left angle of the lower jaw at the masseteric impression, the left mastoid process, the tabular part of the occipital bone, the posterior extremities of the alveolar border of the superior maxilla, the lower end of the right external pterygoid plate, and inner surface of the right angle of muscle. In the spine there is osseous ankylosis of the articular processes throughout its length and of the costo-vertebral joints with one or two exceptions in the upper dorsal region. All the lumbar spinous processes and several of the dorsal spines are united by bone at their extremities. On the posterior aspect of the skeleton two stout columns of bone, one on either side of the midline pass upwards from the pelvis to the shoulder girdle, and are connected in the lumbar region by three transverse bars, which are ankylosed to the spinous processes as they cross the middle line. Starting at the extremity of the sacrum on the left side and at the impression for the gluteus maximus on the right, the two columns are ankylosed by stout osseous masses to the iliac crests, several of the ribs in the mid- and lower dorsal regions, and to the inferior angles of the scapulae; and they terminate superiorly in processes directly forwards towards the neck of the humerus, that on the right side joining a stout column of bone running from the outer end of the spine of the scapula to the deltoid impression of the humerus. From the front of the symphysis pubis and right ischial tuberosity, two thick bars of bone pass downwards to the linea aspera of the right femur, uniting before joining that bone. Small bony outgrowths are present at the margins of the articular surfaces of many of the joints, and ankylosis, apparently due to ossification of ligaments, has occurred extensively in the carpal, tarsal and ankle-joints. The fibulae are ankylosed to the tibiae at both extremities. The left elbow-joint is fixed in a position of semiflexion by a thick bar of bone between the olecranon and the shaft of the humerus immediately above the olecranon fossa. A prominent ridge of bone, resembling an accessory spine, springs from the middle of the left infraspinous fossa, and a similar, smaller process from the right supra-spinous fossa. There is marked thickening of the shafts of the femora immediately below the great trochanter on the posterior aspect of the bones.

Notes by William Clift:

"This skeleton was purchased by Mr. Hunter on November 5th 1783 at the sale of Mr. George Hawkin's collection at Hutchins's Auction Rooms, King Street, Covent Garden, for the sum of 85 guineas."

"Mr. R. Haynes (Hunter's dissecting room attendant) told me he had heard Mr. Hunter or Mr. Bell say that the body had been interred seven years before it could be procured; as is evident from the state of the softer parts of most of the bones." *MS. Catalogue.*

In a partially priced catalogue of this sale, which was purchased at the disposal of Mr Heaviside's collection in 1830, this specimen is listed as Lot 106 —"an anchylosed skeleton". A label pasted inside the box containing the bones stated: "Mr. Jeffs, aged 39".

Localised

P. 805. Sections of part of the shaft of a femur, near the centre of which on the outer aspect is a flat bony outgrowth 4 inches in length, the inner margin of which is sharp and overhangs the anterior surface of the shaft.

P. 806. A left femur from the upper half of the shaft of which a stout plate of bone, 5 inches in length and $1\frac{1}{2}$ inches in breadth at its lowest part, projects downwards and forwards from the anterior surface of the shaft. Its outer border has a linear attachment to the bone about 3 inches long, while the inner border is free and narrow.

17. *Primary Tumours*
P. 807-840

Skull—Compact Osteoma

P. 807. The upper portion of a frontal bone. Growing from the anterior surface near the middle line there are two small lenticular tumours of extremely compact osseous tissue.

The surface of each is highly polished, and each slightly overlaps its base of attachment. Of the lesser tumour, one half has been cleanly detached from the outer table.

P. 808. Parts of a frontal and of a parietal bone, on the inner surface of which are several osseous growths. These growths are fixed on the inner table of the skull by broad bases; their exposed surfaces are irregularly nodulated. The dura mater is extended over them, and attenuated; in some parts, an attempt has been made to reflect it from them; in other parts recent lymph is deposited upon the inner surface of the dura mater. The largest of the osseous growths is beneath the coronal suture; it has an irregularly oval form, measuring about 2 inches in length, 1 inch in breadth and at the most one-third of an inch in thickness: the other growths are much smaller in all their dimensions. The texture of the bones of the skull appears natural; but in the temporal region there is a considerable increase of thickness. A trephine has been applied at the part around which the chief deposit of lymph on the inner surface of the dura mater has taken place.

P. 809. The anterior portion of a skull, in the frontal sinus of which a compact osteoma has grown. The tumour is of an irregular rounded form, is deeply lobed and nodulated on its surface, and measures about 6·3 cm. ($2\frac{1}{2}$ inches) in its greatest diameter. It has for the most part the hardness and apparent structure of ivory but in its central and posterior part is composed of very close cancellous tissue.

In its growth, the tumour has perforated the anterior wall of the frontal sinus, as well as the roof of the left orbit and the posterior wall of the sinus, in which position it forms a prominent lobulated elevation within the cranial cavity. In all the situations shown it is unconnected with the frontal bone, its attachment being evidently of very localised extent.

On the frontal bone, a short way behind the growth, there is a second compact osteoma of lenticular form.

Hunterian Drawing No. 108. Baillie's *Morbid Anatomy*, Fasc. x, pl. 1, fig. 2. Everard Home, *Philosophical Transactions*, 1799, vol. lxxxix, p. 239.

P. 810. A calvaria. From the posterior edge of the frontal bone on the right side and near the temporal ridge, there has grown a small osteoma, about the size of a haricot. Posteriorly it overhangs its base of attachment.

The divided surface presents a structure like, and continuous with, the diploë, closed in with a layer of compact tissue which is continuous with the outer table.

Finger—Cancellous Osteoma

P. 811. The first two phalanges of a finger. The proximal phalanx is greatly expanded by the growth of a tumour within the shaft. In two situations the tumour has penetrated the compact tissue of the shaft, forming cauliflower-like excrescences on the palmar aspect, consisting of cancellous bone.

Femur—Cancellous Osteoma

P. 812. Part of a tibia, with a small round flat osseous tumour, consisting of cancellous tissue, on the middle of its anterior surface.

Humerus—Multiple Exostosis

P. 813. A right humerus from a young person. The upper part of the shaft is flattened laterally and presents on its anterior aspect two, and on its posterior, three, small bony projections.

Radius and Ulna

P. 814. A radius and ulna, without their epiphyses, from the same case as the preceding. They are of very light, brittle texture, atrophied and curved. On the inner margin of the lower end of the radius is a spinous cylindriform growth of bone, with a broad expanding base, cancellous in its interior, but externally, except at its extremity, covered with a thin layer of compact bone continuous with the wall of the shaft. The middle of the outer margin of the ulna is prolonged in a thin broad ridge, the free border of which is unequally thickened and rests in a deep pit in the side of the radius. There is also a small pointed growth, like that on the radius, just beneath the coronoid process of the ulna.

Femur

P. 815. The right femur from the same case as the preceding, on which a small sharp-edged ridge projects in front of the lesser trochanter. There is also a rounded bony projection at the lower part of the popliteal surface.

P. 816. The left femur from the same case as the preceding specimen, on which a ridge of cancellous bone projects on the outer margin, just above the external condyle. There is a similar but smaller projection on the popliteal surface.

Tibia

P. 817. The right tibia from the same case as the preceding specimen, showing a small projection of bone at the upper end of its external border, and three similar projections near its lower extremity.

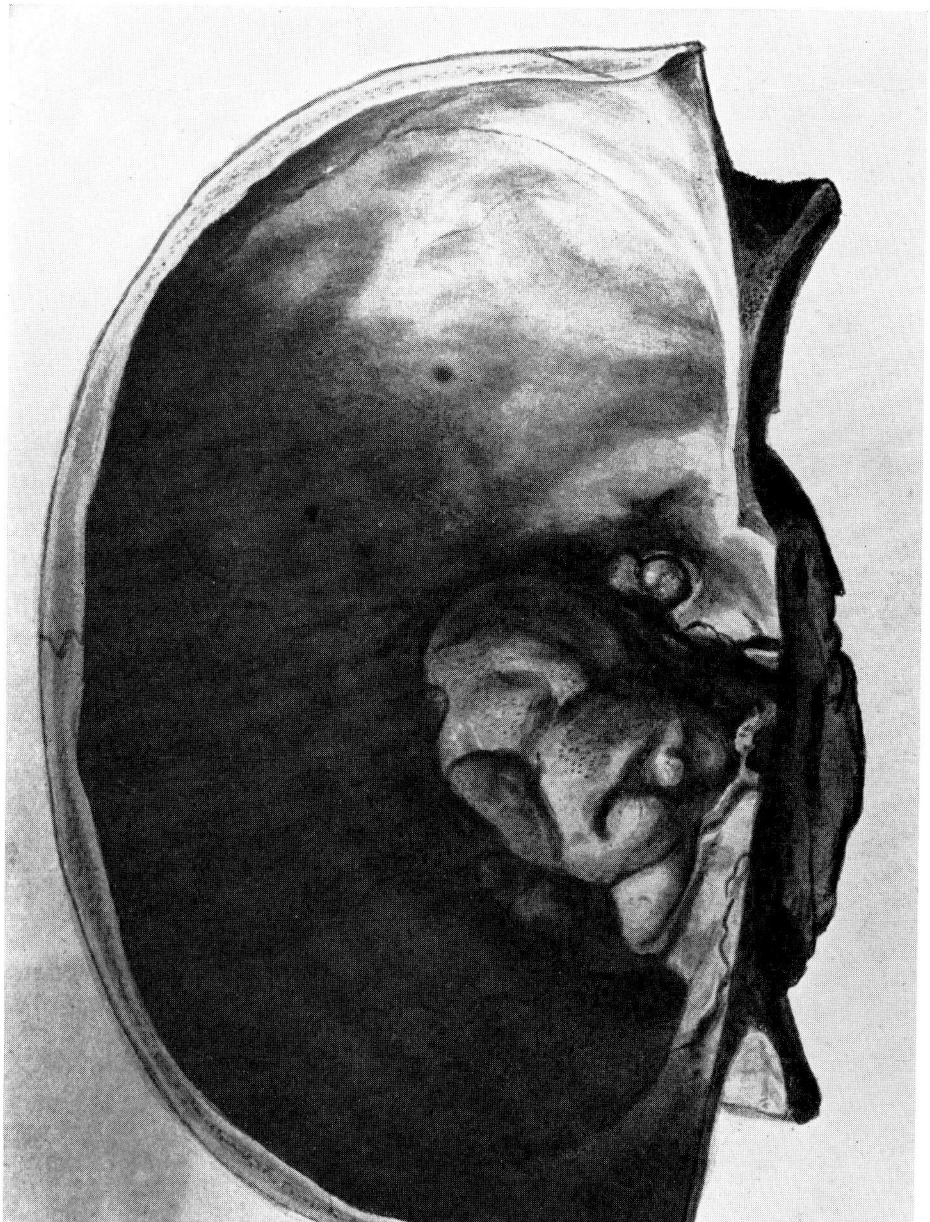

Fig. 9
"The inside of the Os Frontis showing the projection of the Tumour there." Drawing No. 108 by William Bell. Specimen 809.

Face p. 72

FIG. 10
Section of the lower half of a femur with a "bony tumour."
Specimen 826.

Face p. 73

P. 818. The left tibia from the same case as the preceding specimen, on which there are several small pointed outgrowths of cancellous bone, thinly covered with compact tissue, near its upper end, and two of the same kind near its lower end. Of the latter, one seated on the posterior margin of the shaft forms a narrow cylindriform process $\frac{1}{2}$ inch long.

Femur—Exostosis

P. 819. A femur with a small osseous outgrowth upon its inner margin just above the internal condyle. The growth is somewhat flattened, has a broad base, is directed upwards, and is composed of cancellous tissue, the rough, irregular surface of which was probably covered with cartilage.

P. 820. The lower part of a right femur with an osseous outgrowth from the postero-external aspect of the shaft. The growth consists of cancellous tissue with a thin coating of compact bone. It is attached by a broad base, 2 inches in vertical measurement, is flattened at its lower part where it nearly reaches the condyle, and ends superiorly in an irregular projection directed upwards.

Finger—Chrondroma

P. 821. A vertical section of one of the fingers of a child comprising the three phalanges.

The shaft of the second phalanx has been almost destroyed by the growth of a spheroidal chondroma: the other half of this tumour is mounted below the specimen.

In the case of the first phalanx the shaft of the bone and the articular ends are almost completely replaced by a cartilaginous growth, over which the articular cartilage in places alone remains. The general form of the phalanx is maintained, with the exception of some bulging on the dorsal aspect.

The several tumours present the usual lobulated structure.

Scapula

P. 822. "A congeries of encysted tumours which formed between the scapula and ribs of a woman at St. George's Hospital." *MS. Catalogue.*

The growth which is 14 cm. ($5\frac{1}{2}$ inches) in chief diameter, is of somewhat oval form, and consists of a closely combined series of lobules, in which liquefaction has occurred.

Histologically the more intact portions which project into the spaces, consist of cartilage and mucous tissue which merge one into the other.

P. 823. Another tumour in the same situation from the same patient. The growth is soft and friable and contains a number of cystic spaces into which irregular nodular masses project.

Histologically the tumour consists of similar mucoid cartilaginous tissue.

Tibia—Chondrosarcoma

P. 824. A vertical section of a knee, with a tumour of the tibia. The anterior part of the head of the tibia is involved in the base of a large, lobulated, soft, spongy tumour, containing many small cysts, by the growth of which all its tissue seems destroyed. From this position the tumour has extended forwards, stretching the ligamentum patellae and the adjacent parts, and upwards into the cavity of the joint, which it almost completely fills. Its surface within the joint is covered by synovial membrane; but its pressure has caused absorption of part of the cartilages of the femur and patella.

Microscopically the growth is an ossifying chondrosarcoma, showing numerous islands of cartilage, in some of which the cells have undergone proliferation, areas of round and oval sarcoma cells and trabeculae of bone. These structures are separated by broad tracts of fibrous tissue.

Humerus—Periosteal Fibroma

P. 825. A vertical section of the lower half of the upper arm, the section being carried through the humerus. Growing from the soft parts immediately on the front of the shaft there is a large, somewhat hemispherical tumour.

Microscopic examination shows the growth to be composed of intersecting bundles of fibrous tissue; a few foci of finely granular calcareous material are scattered through its substance.

The bone is quite uninvolved; and there is no periosteum distinguishable beneath the neoplasm.

Femur—Osteosarcoma

P. 826. A vertical section of the lower half of a femur, macerated, the inferior portion of the shaft of which, for a distance of 12·7 cm. (5 inches), is surrounded by an osteoid sarcoma.

The medullary canal opposite the external enlargement, is filled with a dense extension of the growth.

A short distance below the site of amputation there is a second hemispherical tumour in the medullary canal, which lies 7·5 cm. (3 inches) above the main growth, the disease having extended within the canal, in this case, for an unusual distance beyond the limits of the external swelling.

The history is as follows:

"Case of a tumour in the thigh-bone, which became bone, as also in the chest.

Bony Tumour in the Thigh

"A man came into St. George's Hospital, November, 1786, with a hard swelling in the lower part of the thigh, as it were beginning from the knee. It appeared to be a thickening of the bone, involving in it the surrounding parts, the whole appearing to make nearly one mass.

"It was then increasing very rapidly, and at this time so large as to interfere with the motion of the joint, so as to render the leg useless. The tumour was hard as bone, and seemed to be nearly equal all round the thigh.

"The man had been in perfect health in this part, till about five months before, and then began to feel shooting or darting pains in that part of the thigh, which continued for four weeks, during which time he could perceive no alteration in the parts, either to appearance or to the feel; but in the fifth week, or beginning of the second month, the part began evidently to enlarge, although at first very slowly; but in the third month it increased more rapidly, and was attended with more pain as it enlarged, so that in the fifth month it was increased very considerably; and the pain was now exhausting him much, increasing nearly in proportion to the size; and it was thought advisable to remove the whole, which was done.

"On examining the diseased part, it was found to consist of a soft substance surrounding the lower part of the thigh-bone of the tumour kind, which seemed to originate from the bone itself, into which were shooting ossifications; and as the tumour formed externally, the ossific matter formed in it, as it were keeping pace with it, the tumour becoming a nidus for bone, similar to a cartilage.

"Besides this external increase of bone, there was the same internally; for the canal as well as the whole cellular texture of the bone was filled entirely up, so that the old bone was become compact and hard in its substance.

"He went on well after the amputation for four weeks, when he began to complain of a difficulty in breathing, but not attended with the least pain. This increased for a fortnight very much, and he then had a rigor every other day; which was supposed to be an attack of ague. From this time he began to lose his flesh and sink gradually, his breathing being more and more difficult, and in three weeks after the difficulty of breathing came on, and seven days after the rigors, he died; living only seven weeks after the operation.

Dissection

"On examining the body it was found that the same disposition to form bone, which had taken place in the thigh-bone, had also affected the thorax and its contents.

"The cartilages of the trachea were opaque and in some places ossified; bony tumours were found in the cellular membrane of the lungs, upon the pericardium, and some very large ones on the pleura, adhering to the ribs, and upon the anterior surface of the vertebrae of the back.

"These tumours, or diseased appearances, were in all the different states between soft parts and bones; so that they had not been originally formed into the bone, but had been formed first of soft substance, which had afterwards been removed for osseous matter.

"Besides the singularity of the disease and appearances themselves, the seeming quick progress of the disease in the chest is not less uncommon or unaccountable; for when the leg was amputated he had not the least symptom of any disease in the chest, nor for four weeks after; and in three weeks after the first symptoms he died, in which time (from the symptoms) we are to conclude these tumours grew. However, from reasoning, we may suppose that the appearances which were found after death had taken place a considerable time before the symptoms took place: yet we must allow that they could not have advanced far: and therefore their growth must have been very rapid, as indeed the increase of the symptoms would plainly show; and if we compare this with the increase of the thigh-bone, which was visible, we can also judge of their increase; for from the first appearance of increase, to his coming into the Hospital, was only between three and four months, and it increased more the last month than all the others.

"One can figure to themselves a reason why the tumour which formed on the outer surface of the thigh-bone might become bony, because it might acquire that disposition from the bone it surrounded; but, from these tumours formed in the chest becoming bone, shows it was the nature of the tumours themselves." *MS. Cases and Dissections*, No. 78.

P. 827. Half of the lower end of the femur, macerated, from the same case as the preceding specimen.

The lower end for a distance of 11 cm. ($4\frac{1}{2}$ inches) is enveloped in a large osteoid sarcoma which projects, however, mainly from the posterior aspect. The growth has extended into the cancellous tissue and within the medullary canal, but especially in an unusual manner, beyond the lower limit of the femur, behind the knee-joint, where it reaches for a distance of 7·5 cm. (3 inches).

As seen at the back of the specimen, the tumour is deeply grooved and perforated for the passage of several anatomical structures, the identification of which is not recorded.

P. 828. Another section of the same femur.

P. 829. The lower half of a femur, surrounded and filled with a large, heavy, and very irregular osteosarcoma, which bears a close resemblance to the preceding tumours.

P. 830. Sections of a hard and heavy osteosarcoma. It is not known from what part it was removed.

Rib

P. 831. The rib of a Horse, surrounded through nearly its whole length with a thick irregular growth of light, porous bone, like the osseous part of an osteosarcomatous tumour. The substance of the rib is involved in the growth; and at one end its outer surface seems to be expanded into that of the tumour. On the surface of the tumour there are numerous large round and oval apertures leading directly, or by sinuous canals, into its interior.

Horn of Ox

P. 832. The osseous matrix of the horn of an Ox, on the extremity of which is a large spongy mass of bone such as is met with in osteosarcomata. Its substance is light, porous and composed of slender, osseous spicules radiating from the end of the matrix, with which its tissue has completely coalesced. In places the surface is marked by deep grooves for blood-vessels. The specimen has been divided into three pieces.

Jaw of Opossum

P. 833. The lower jaw of a Virginian Opossum, on which there is at the anterior part of the left half an oval tumour of sponge-like osseous tissue, composed of spicules of bone arranged vertically to the surface of the jaw. The growth is probably an osteosarcoma.

Tibia—Osteoclastoma

P. 834. "A section of a tumour, formed upon (or rather in) the head of the tibia which gave an appearance of immense size to the joint of the knee. Besides the general volume given to the knee, there was a protuberance of the size of a large egg, which is become smooth, and of a bluish hue from the contents shining through, which is seen in the preparation. The leg was removed, and upon examination it (the tumour) was found to have arisen on the head of the tibia. It consisted of a great many cavities or cells, filled with bloody serum, and even coagulated blood of long standing." *MS. Catalogue.*

The tumour has undergone extensive cystic degeneration, the spaces of which are filled with blood. The tissue which intervenes between the cystic spaces is soft and discoloured by haemorrhage. Owing to the expansion of the upper extremity of the tibia by the growth, no bone is recognisable in the section. Microscopic examination shows the neoplasm to be a sarcoma, in places formed of round, in others of spindle cells, with which are many large multinucleated cells.

P. 835. "Is the joint of the knee, with the head of the tibia, in which the before-mentioned tumour was formed, showing also where it is pushing into the cavity of the joint." *MS. Catalogue.*

The cartilage of the tibia, in the internal fossa, is elevated by the tumour beneath it, but the growth has not penetrated the cartilage. The head of the bone is expanded, but osseous tissue can be recognised.

P. 836. The upper end of a tibia. For a distance of 10 cm. (4 inches) the bone has been destroyed by the growth of a central tumour, probably a giant-celled sarcoma, a characteristic shell of new bone having been produced in connection with the periosteum.

The osseous shell is hardly anywhere complete, but presents large numbers of thinly edged, irregular gaps of varying size; from the interior there project a certain number of sharp ridges or stunted septa. The layer of articular bone has not been destroyed, the form of the articular surfaces being still intact.

Femur

P. 837. The lower part of a femur, the end of which is expanded owing to the presence of a central tumour. The resulting cavity which measures 5 inches in diameter is incomplete at its posterior part; its walls vary from half a line to $\frac{1}{4}$ inch in thickness; their exterior is smooth, but their interior is rendered very irregular by ridges forming imperfect partitions across its cavity. The articular surfaces of the condyles are ulcerated. The shaft of the femur ends abruptly at the upper part of the cavity.

Tibia

P. 838. The upper part of a tibia which has become expanded owing to the presence of a central tumour. The resulting cavity measures $5\frac{1}{2}$ inches in diameter and is bounded by an incomplete bony shell which is widely open anteriorly and elsewhere is perforated by large apertures. Internally the cavity is marked by incomplete bony septa, which separated the lobules of the growth.

P. 839. The upper part of a tibia which has been expanded by a central growth. The resulting cavity is $3\frac{1}{2}$ inches in diameter and is surrounded by a thin fibrous wall in which thin plates of bone are embedded. The articular cartilage is normal.

In Hunter's time this condition was known as spina ventosa, i.e. enlargement and thinning of a bone which is the seat of a neoplasm or caries.

Femur—Unidentified Tumour

P. 840. A section of a knee-joint. The lower extremity of the femur is the seat of a large, friable growth which has destroyed the whole thickness of the bone, separating the lower extremity from the shaft, and projects anteriorly under the crureus muscle and posteriorly fills the popliteal space. The knee joint is healthy.

Histologically the growth is a round-celled sarcoma, which contains a certain number of giant cells.

18. *Secondary Tumours*
P. 841-850

Skull

P. 841. Part of the parietal bone and overlying scalp. A tumour measuring about $1\frac{1}{2}$ inches in diameter is situated in the substance of the bone and forms a rounded projection on its inner surface. The growth is adherent and infiltrates the dura mater, which shows several separate nodules in its substance.

Microscopic examination shows that the neoplasm is a spheroidal-celled carcinoma; it is obviously a metastasis, probably secondary to carcinoma of the breast.

P. 842. The other section of the same parietal bone and tumour.

P. 843. Part of a parietal bone to the inner surface of which a small circular flattened tumour is attached. The growth is somewhat spongy in texture, and does not appear to have been adherent to the dura mater. Histological examination shows that the growth is a spheroidal-celled carcinoma, and must have been a metastatis from a primary tumour in some other part of the body.

P. 844. A calvaria which, owing to the presence of secondary malignant growths in its substance, has been perforated in several situations, and has numerous deep circular ulcerated cavities, with sharp broken margins, in its internal table and diploë. The disease is most marked along the course of the longitudinal sinus, near which, besides presenting several deep excavations from within, and a perforation an inch wide at the angle of the occipital bone, the substance

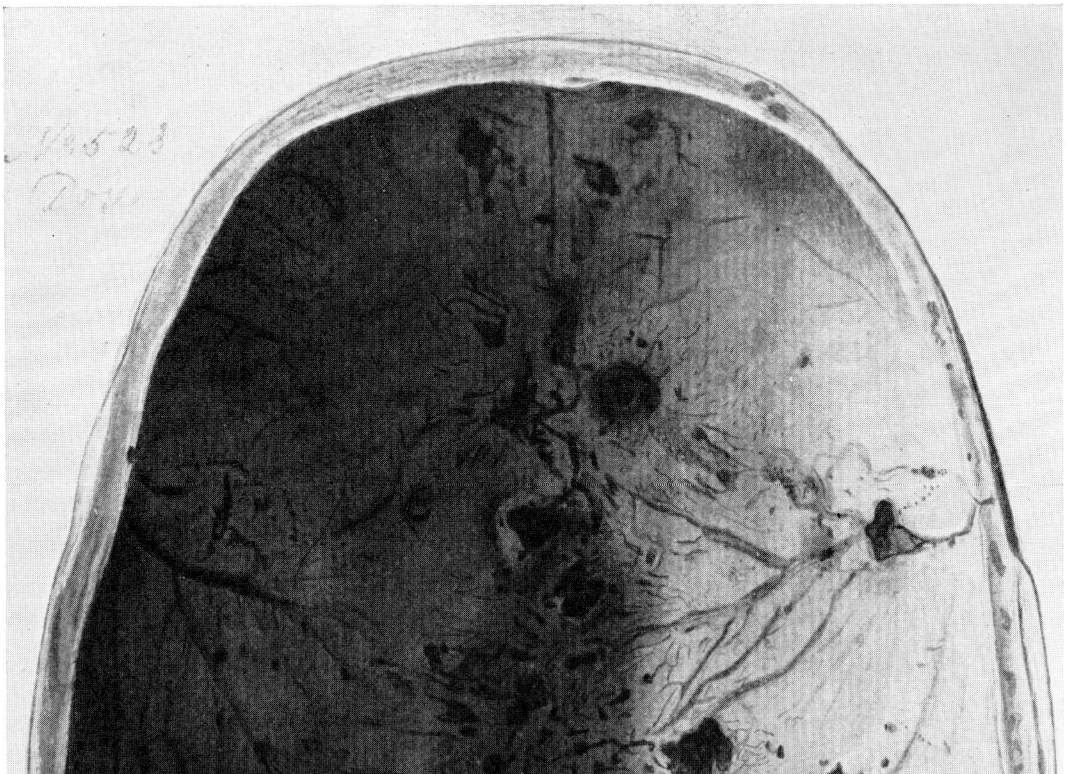

FIG. 11

"Is a view of the inside of the skull where the ulceration seems to have begun and in some parts gone quite through the skull." Drawing No. 43 by William Bell. Specimen 844.

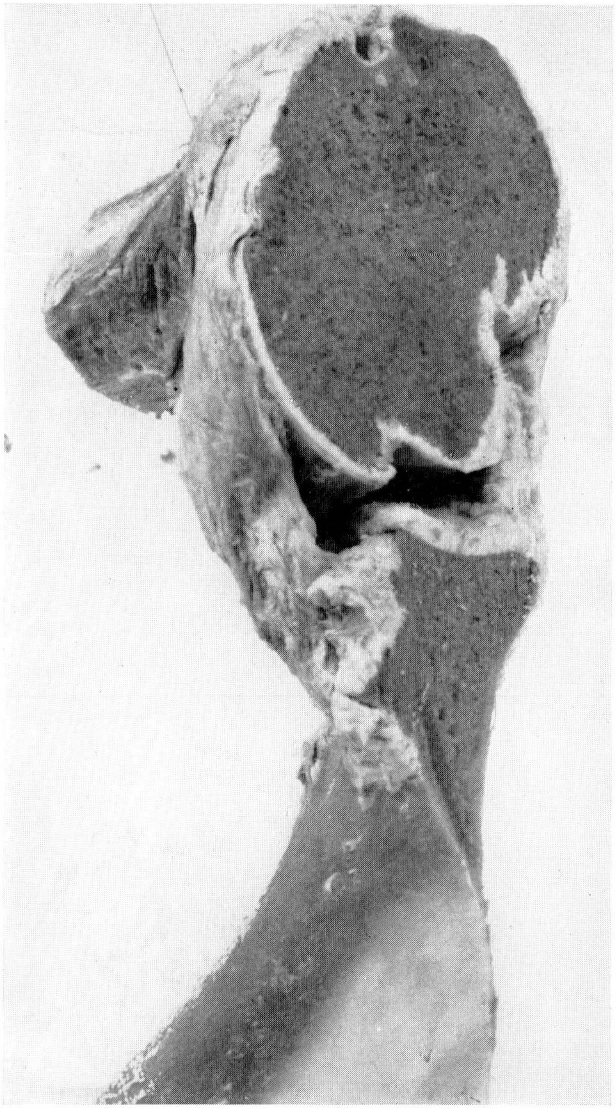

Fig. 12
Section of a shoulder-joint showing an old incompletely reduced subcoracoid dislocation. Specimen 858.

of the skull is riddled with numerous small round apertures, of which some open internally, some externally, and others pass completely through the bone. Around the margins of the larger apertures and cavities small quantities of new bone are deposited on the inner table. Spontaneous fracture of some of the bones occurred during life.

"From the Hon. George Grenville.—On opening the body, many other bones were found in the same condition, where, in place of the bone that was removed, there was found a curdly substance." *MS. Catalogue.*

The Rt. Hon. George Granville died on November 13th 1770 aged 58.
Drawing No. 43.

Rib

P. 845. Portion of a rib from the same case as the preceding specimen. The bone has been the seat of a central growth, leading to absorption of the compact layer in places. Drawing No. 44, fig. 3.

Humerus

P. 846. The lower end of the left humerus of the same case as the preceding. The substance of the shaft, through a length of nearly 3 inches, has been extensively destroyed by a central malignant growth. The remaining portion of the wall fractured under a very slight force, a short time before death.
Drawing No. 43, figs. 1, 2.

Femur

P. 847. The upper half of the right femur from the same case as the preceding. The upper part of the shaft has been slightly expanded by a central growth, the resulting cavity being surrounded by a thin shell of bone, the surface of which is covered with a deposit of new bone. The bone is said to have fractured under a slight force.

Skull

P. 848. A calvaria showing in the left posterior parietal region a perforation $\frac{2}{3}$ inch in width, around which a narrow foliated deposit of new bone projects above the surface of the skull: on the inner surface there is less regular formation of bone around the aperture and extending for a considerable distance beyond it. Internally, in the frontal region there is a rounded ulcerated area, with a broad zone of osteophytes around it; and on the inner surface of the occipital bone is a similarly ulcerated surface bounded in part by osteophytes. In several other situations there are small carious foci, which in some have led to perforation, and on the inner surface of the parietal bones shallow ulcers are present associated with a varying amount of new bone formation, and porosity of the bone. The whole of the external surface of the calvaria is marked by closely placed small circular apertures, probably occupied by dilated vessels; but over considerable areas in the parietal and frontal bones there also appears to have occurred a very superficial erosion. It is suggested that the calvaria was the seat of secondary growths but the precise nature of the lesion is doubtful.

The specimen was originally marked "September 3, 1779. Died about 5 weeks ago, aged 19. G. Brande"; and in Hunter's handwriting, "supposed to be venereal". Drawings Nos. 41, 42.

P. 849. A calvaria, in which there is in the frontal bone a circular perforation $2\frac{1}{2}$ inches in diameter, with abrupt jagged edges. The aperture is closed by a thin fibrous membrane, containing thin plates of bone in places, which has probably resulted from atrophy of the external table owing to pressure of a subjacent growth. On the interior of the rest of the skull there are several deep ulcerated cavities, with sharp abrupt margins, which are for the most part circular in form and from two to six lines in diameter. They extend through the inner table and diploë; the external table over them is very thin, and in a few situations is perforated by small apertures. The disease was probably consequent on the growth of malignant tumours.

P. 850. Portion of a skull, removed with a trephine. There is a perforation in it, such as may have been produced by a tumour growing from within outwards.

19. *Altered by the Growth of Tumours*
P. 851-855

Jaw of Sheep—Cyst

P. 851. The right half of the lower jaw of a Sheep the anterior part of which is dilated into a cyst of regularly oval form, measuring $3\frac{1}{2}$ inches in its longest diameter. There is a round aperture about 1 inch in width at the anterior part of the cyst, which probably had its origin in connection with the teeth.

Jaw—Eroded by Tumour

P. 852. A lower jaw of which the alveolar border of the symphysis and right half of the body has been destroyed, probably by the growth of a tumour. There is a scanty deposit of new bone at the margins of the affected area.

Scapula

P. 853. A left scapula in which the glenoid cavity, acromion, and coracoid process have been destroyed by the growth of a large tumour.

Clavicle

P. 854. A clavicle the shaft of which is extensively eroded, probably by a new growth. The outer extremity is abnormally light and porous, and on both extremities new bone has been deposited.

Tibia and Fibula

P. 855. The lower halves of the tibia and fibula of a young subject. The lower three inches of the shaft of the tibia and the greater part of its lower extremity has been destroyed by a malignant growth. The upper limit of the growth is indicated by a deposit of new subperiosteal bone which forms an expansion over the remnant of the shaft which was embedded in the tumour. Part of the epiphyseal cartilage is still intact. There is a deposit of new bone on the fibula adjacent to the tumour. The ankle joint is healthy.

II

Joints

1. *Dislocation*

P. 856-867

Normal Shoulder

P. 856. A section of a healthy shoulder-joint, for comparison with the effects of injury and disease. The section is made in a horizontal plane below the spine of the scapula, and through the middle of the glenoid cavity and the head of the humerus.

P. 857. The other section of the same joint.

Shoulder—Dislocation

P. 858. Section of a shoulder-joint, cut in a horizontal plane through the middle of the glenoid cavity, showing an old incompletely reduced subcoracoid dislocation. The articular surfaces are only partially adapted; the anterior two-thirds of that of the humerus are not applied against the glenoid cavity, and are separated by a deep depression from the posterior third which is so applied. The articular cartilages appear healthy; the capsule is shortened and very much thickened. The head of the humerus is in contact with the under surface of the coracoid process.

P. 859. The other section of the same joint.

P. 860. A left shoulder-joint in which there has been a subcoracoid dislocation of the humerus. There is a rent, an inch long, in the anterior and inner part of the capsule (the larger opening in the upper part of the capsule is artificial). The glenoid fossa is covered with a thin layer of fibrous tissue and the synovial membrane is slightly thickened and has numerous pedunculated villous processes attached to its upper part. There is slight irregularity and lipping of the articular surface of the humerus, due to osteoarthritis.

P. 861. A shoulder-joint with an old unreduced subcoracoid dislocation of the humerus. The head of the humerus rests below, and in front of, the glenoid cavity, in a hollow bounded by a deposit of bone, on the anterior margin of the glenoid cavity and on the belly of the scapula. The surface of the fossa is in part bony, hard, and smooth, and in part covered with fibrous tissue; the posterior portion of the head of the humerus, which rested and moved in it, is flattened, and its articular cartilage appears of natural thickness. The rest of the cartilage is irregular, lipped, and in parts destroyed, owing to osteoarthritis. The upper and posterior part of the capsular ligament is torn; the margins of the rent, indicated by bristles, are thickened and beset with a fringe of fibrous pedunculated growths. The greater tuberosity has been broken off and reunited.

P. 862. The bones of a left shoulder-joint. The head of the humerus has been dislocated forwards beneath the coracoid process and rests on a shallow concave surface of new bone formed on the scapula. A plate of bone thrown out from the shaft of the humerus immediately below the great tuberosity articulates with the glenoid cavity. The tip of the acromion has been fractured and is united after slight displacement.

Elbow

P. 863. The lower end of a right humerus, with the radius and ulna, showing a congenital dislocation backwards of the head of the radius. The head of the radius is small; it lies below the external condylar process in the flexed position of the joint and its anterior part and the adjacent part of the neck articulates with the lesser sigmoid cavity of the ulna. The upper surface of the olecranon and the head of the radius are almost at the same level and the styloid process of the radius is on a level with that of the ulna. The ulna articulates in the normal way with the humerus, except that the articular surface is expanded laterally and rests partly against the capitellum. The shafts of both radius and ulna are altered in shape; the upper part of the ulna is unduly curved and anteriorly presents a sharp crest; the shaft of the radius is flattened.

Metacarpal

P. 864. The bones of a left hand. The carpal end of the fourth metacarpal bone has been partially dislocated backwards, and half its articular surface has lost its relation to the unciform bone. Its radial margin is ankylosed by bone to the adjacent margin of the third metacarpal.

Hip

P. 865. Section of a hip-joint, in which the femur was dislocated backwards towards the sciatic notch and not reduced. The section was made in a horizontal plane through the lower part of the ilium, an inch above the spine of the ischium, and through the head and neck of the femur and the great trochanter. The head of the femur rests on the back of the ischium, between the posterior margin of the acetabulum and the ischial spine. The inferior and inner surface of the head and neck is united by ligamentous tissue to the surface of the ischium. A capacious and thick capsule encloses the dislocated head and neck. The acetabulum is reduced in size and filled with a tough ligamentous tissue. The cartilage on that part of the head of the femur which is not adherent to the ischium is thin, and its surface uneven. The femur is markedly flexed and adducted, and a process of bone developed on the antero-internal aspect of the shaft articulates with a smooth oval surface on the anterior margin of the acetabulum.

P. 866. The upper section of the same hip joint, exhibiting more distinctly the layer of ligamentous tissue uniting the inner surface of the head and neck of the femur to the outer surface of the ischium, and showing that the end of the round ligament, which was torn away from the bottom of the acetabulum, has become firmly adherent to the inner surface of the capsule, near its attachment to the posterior and inferior spine of the ilium. The section of the head of the femur shows considerable fatty atrophy of the cancellous tissue in the part not in contact with the hip-bone.

Ankle

P. 867. The bones of an ankle-joint, in which the tibia is stated to have been long dislocated. Nearly all the articular cartilages have been destroyed, and the bony surfaces are intact, except for a small ulcerated area at the under part of the head of the astragalus. A few fibrous adhesions unite the articular surfaces of the tibia and astragalus.

2. *Pyogenic Arthritis*
P. 868-882

Wrist

P. 868. The right semilunar and unciform bones, with the os magnum. The cartilages of the former bones are extensively ulcerated, the margins of the ulcers being sharply cut, the exposed compact layer of bone destroyed and the cancellous tissue in places rarefied. The outer articular surface of the os magnum is covered with a layer of fibrinous exudate.

P. 869. A right radius showing the effects of suppuration of the wrist-joint. The compact layer of the articular surface of the lower extremity of the bone has been evenly destroyed by ulceration, exposing the unaltered cancellous tissue beneath. The surrounding external surface of the extremity is covered with osteophytes.

Finger

P. 870. A finger, with part of its metacarpal bone. The articular cartilages have been to a large extent destroyed and the subjacent bone is eroded in places. Numerous fibrous adhesions run between the articular surfaces. The changes are probably due to acute inflammation of the joint.

Hip

P. 871. The bones of a right hip-joint showing changes due to suppurative inflammation. The articular surface of the acetabulum has been almost entirely destroyed and the cavity deepened by ulceration. Several perforations are present, the upper and largest of which opens the into lower part of the iliac fossa where the limits of an abscess are marked by a deposit of new bone. A somewhat smaller perforation opens into the pelvic cavity. The articular surface of the head of the femur is denuded of its compact layer, the subjacent cancellous tissus being but little altered. Drawing No. 39.

P. 872. A left hip-joint, with the upper part of the femur. The floor of the acetabulum is extensively eroded, probably by suppuration, and a large part of the anterior margin and central part of the floor completely destroyed. A deposit of new bone has occurred within the pelvis at the lower margin of the perforation and marking out the lower limit of an area of superficial rarefaction extending from the ischial spine to the obturator groove, corresponding to the site of a pelvic abscess. The joint is also affected with chronic osteoarthritis. The head of the femur is deformed, somewhat conical in shape and depressed below its normal level; its articular surface is worn away in one situation, exposing the pitted and atrophic cancellous tissue, and its margin has bony outgrowths in places. The neck and upper part of the shaft exhibits an abnormal flattening from before back.

Knee

P. 873. "The cavity of the knee-joint, very much enlarged and its internal surface inflamed." *MS. Catalogue.*

A vertical section of a knee-joint. The articular cartilage of the femur and tibia has been to a large extent destroyed and the exposed bone on the condylar surface of the femur is covered with granulation tissue. The supra-patellar pouch is greatly distended and its slightly thickened synovial lining is covered with fibrino-purulent exudate.

P. 874. A patella the articular surface of which has been destroyed, probably by suppurative arthritis, exposing the almost unaltered cancellous tissue.

P. 875. A patella, the articular cartilage of which has been completely destroyed, exposing the subjacent cancellous tissue, which is unaltered. The condition has probably resulted from acute suppuration.

P. 876. A vertical section of a knee-joint. The articular cartilage is extensively thinned and eroded. The supra-patellar pouch and joint cavity are slightly distended and the fibrous capsule and synovial membrane slightly thickened. The inner surface of the synovial membrane and part of the articular surface of the femur are covered with a fibrino-purulent deposit.

P. 877. A vertical section of the lower end of a femur and of some of the adjacent tissues, from the knee-joint last described. There is a perforation in the lateral wall of the distended subcrureal pouch, communicating with a small abscess cavity, outside the joint.

P. 878. A vertical section from the same knee-joint.

P. 879. The lower extremity of a tibia, from the articular surface of which a plate of bone has been exfoliated. The surrounding cartilage has been removed by ulceration, and the subjacent bone appears to be replaced by granulation tissue.

Ankle

P. 880. The lower end of a tibia with part of an astragalus. The upper half of the astragalus has been removed by ulceration; its remaining portion, dead and deprived of its articular cartilages, lay loose in the ankle joint. The cartilage of the articular surface of the tibia has been destroyed and replaced by a layer of granulation tissue.

P. 881. An astragalus, from the upper and lateral surfaces of which the anterior half of the articular cartilage has been destroyed apparently as the result of acute arthritis. The exposed bone is healthy. The mucous membrane is, in places and more particularly at the posterior part of the joint, covered with a layer of fibrinous exudate. Similar changes are seen on the under surface of the bone, in connection with the astragalo-calcaneal articulation.

Foot

P. 882. Parts of a tarsus and metatarsus. The scaphoid and one of the cuneiform bones, deprived of articular cartilage and dead, lie loose in a cavity the interior of which is covered with very vascular granulations. A large portion of the scaphoid bone has been removed by ulceration. Another of the cuneiform bones has also separated after necrosis, and lies loose in a cavity near the sole.

3. *Tuberculosis*
P. 883-920

Spine

P. 883. Vertical sections of the skeleton of the trunk of a child. The bodies of the sixth and seventh dorsal vertebrae are wholly, and those of the fifth, and eighth are partially, destroyed by caries. There appears to have been very little approximation of the vertebral bodies, and the resulting angular deformity is slight.

P. 884. A vertical section of the lumbar and part of the dorsal portion of a child's vertebral column, with the spinal cord. The bodies of the twelfth dorsal and first lumbar vertebrae with the intervertebral cartilages have been completely destroyed, and those of the eleventh dorsal and second lumbar to a large extent destroyed by tubercular caries. The posterior part of the third lumbar vertebra is excavated by caries and the posterior common ligament separated from the bone by an abscess cavity, which is continuous superiorly with a thick-walled abscess-cavity extending backwards to the left transverse process of the first two lumbar vertebrae. There is a marked angular curvature opposite the diseased bone, but the spinal canal is not reduced in width.

P. 885. The skeleton of the trunk of a person about 18 years old, divided by a vertical section through the spine and sternum. After almost complete destruction by ulceration of the bodies of the last five dorsal and the first lumbar vertebrae, the remaining parts have come together and united in one mass of cancellous tissue. There is a very acute angular curve, and a slight lateral distortion, in this region; but the spinal canal is not narrowed. The ribs are approximated, many of the intercostal spaces being only ¼ inch wide; the obtuseness of the angles of the ribs is diminished; and the sides of the chest are flattened, and the extremities of the lower ribs are nearly in contact with the crests of the ilia.

P. 886. The heart and large vessels, injected and dried, from the same case as the preceding specimen. With the heart held in the normal position the descending aorta is seen to pass backwards and downwards for about 4 inches and then bends at an acute angle and passes directly downwards for 2 inches, becoming suddenly greatly diminished in diameter and after making a second acute bend terminates about 3 inches below. The deformity of the aorta is related to the curvature of the spine.

P. 887. The lumbar vertebrae and pelvis of a child. The bodies to the last two lumbar vertebrae are almost entirely destroyed by caries: those of the second and third are very deeply ulcerated; and there has been superficial ulceration of the whole of the front of the sacrum. There is a well-marked angular curvature with its apex at the fifth lumbar spine.

P. 888. Three lumbar vertebrae and a sacrum. The adjacent surfaces of the two lower lumbar vertebrae have been eroded by tubercular caries, the fifth lumbar having a deep excavation towards the spinal canal. The surrounding anterior surfaces of the bones are covered with osteophytes, which doubtless marked the site of an abscess. All the bones are abnormally light and porous.

P. 889. A sacrum with the last three lumbar vertebrae. The anterior surfaces of the last lumbar and first sacral vertebrae are superficially eroded by tubercular caries, and the left lateral mass of the sacrum is similarly and more deeply affected. The adjacent bone is irregular from osteophytes, marking the limits of an abscess, which extended inferiorly to the junction of the second and third sacral bodies.

P. 890. Three lumbar vertebrae, the bodies of which have been irregularly ulcerated by tubercular caries. Both surfaces of the upper vertebra and the adjacent surfaces of the lower show depressions caused by caries, exposing the cancellous tissue which appears unaltered. The osseous tissue of all the bones is lighter and more porous than normal, and on their anterior surfaces there is a slight deposit of new bone. A slight lateral curvature has resulted from the excavation on the left side of the body of the lowest vertebra.

P. 891. The two lower dorsal and the lumbar vertebrae of a young subject. Their tissue is very light, and spongy; and there are scarcely any remains of the compact layer which normally covers the surface. The greater part of the left side of the arch, spinous process, and inferior articular process of the second lumbar vertebra has been destroyed by caries; and deposits of new bone on the sides of the bodies of the adjacent vertebrae make it probable that there was a lumbar abscess.

Shoulder

P. 892. The bones of a right shoulder-joint showing the results of tuberculous disease.

The head of the humerus is completely destroyed and the tuberosities are excavated by a large cavity probably resulting from necrosis of the cancellous tissue. The surrounding surface of the bone is thickly covered with osteophytes. The scapula exhibits a similar excavation of the floor of the glenoid fossa, with an abundant formation of osteophytes on the surrounding bone.

Elbow

P. 893. The bones of an elbow-joint, of which the corresponding articular ends have been extensively destroyed, probably by tuberculous disease. Large excavations are present in both the humerus and ulna; and of the articular portion of the latter bone hardly more than a thin shell remains. The outer part of the head of the radius is partly detached as a sequestrum. New bone has been formed on the surfaces of the shafts, doubtless as the result of associated suppuration.

Finger

P. 894. A finger, in section, swollen by a deposit of tubercular granulation tissue connected with the distal end of the first phalanx. The palmar surface of the bone is superficially ulcerated by the deposit. The disease probably arose in the periosteum of the phalanx. Microscopically the deposit consists of tuberculous granulation tissue.

Hip

P. 895. The right hip-joint of a young subject. The acetabulum is deeply excavated by caries, probably of a tuberculous origin. The floor of the cavity is perforated by an aperture, about $\frac{3}{4}$-inch in diameter, partly filled with granulation tissue and containing a small sequestrum. The cartilage of the head of the femur has been destroyed and the subjacent bone superficially ulcerated. The posterior part of the head and adjoining part of the neck are flattened. Owing to the destruction of the upper margin of the acetabulum, the head of the femur was dislocated upwards and rested on the thick pad of fibrous tissue which adheres to the dorsum ilii and forms part of the thickened capsule of the joint. The synovial membrane is thickened in places, more especially at the hinder and lower part of the neck of the femur.

P. 896. The right hip-joint of a child. The Y-cartilage has been entirely destroyed at its central part, together with the whole of the articular cartilage of the acetabulum, and the subjacent bone is irregularly pitted and ulcerated by caries. The acetabulum is much enlarged owing to the destruction of its margin. The upper extremity of the femur is greatly deformed, almost the whole of the epiphysis of the head having been destroyed by caries. The exposed cancellous tissue is filled with caseating granulations. The ligamentum teres is destroyed and the capsule of the joint thickened.

P. 897. A left hip-bone, with the upper end of the femur, showing changes probably due to tuberculous disease of the hip-joint. The head of the femur and the floor of the acetabulum have been largely destroyed by caries and are very irregular in form. The exposed cancellous tissue is in most places smoothly healed over and the cancellous spaces, especially in the floor of the acetabulum, partly or completely closed by bone. At the junction of the head with the anterior part of the neck of the femur is an elongated, sinuous smooth surface which articulated with the upper and anterior part of the acetabulum, the femur being markedly flexed, in which position it was possibly fixed by fibrous ankylosis. A stout plate of bone has been formed in the region of the small trochanter and a smaller outgrowth springs from the ilio-pectineal eminence; these have probably been deposited in connection with the ilio-psoas muscle.

P. 898. Part of a left hip-bone, with the upper end of the femur, probably affected with tuberculous disease. The floor of the acetabulum is irregularly and deeply excavated by caries and several masses of cancellous tissue appear to be in process of separation as sequestra. It is perforated in two places, one towards the pelvic cavity, the other towards the great sciatic notch. The head and anterior part of the neck of the femur are deeply ulcerated and the whole surface of the bone is rarefied, as is also the ilium above and behind the margin of the acetabulum.

Knee

P. 899. "A knee-joint nearly obliterated by soft union." *MS. Catalogue.*

The anterior part of the capsule with the patella reflected downwards. The synovial membrane is thickened and pulpy and the fibrous capsule thickened. The articular surface of the patella is covered with pulpy granulation tissue. The cartilage of the tibia is to a great extent destroyed and in its outer half is a perforation leading into a diseased focus in the cancellous tissue. The articular cartilage of the femur is in parts covered with granulation tissue, in parts superficially eroded; its margins are deeply eroded and the subjacent bone excavated and containing small sequestra. In the skin over the patella and in the base of an ulcer on the lateral aspect of the knee are several sinuses.

P. 900. The upper part of a tibia, in the outer side of the head of which is a large carious cavity with a widely open orifice. The knee-joint communicates with a part of the cavity; and the articular surfaces are superficially ulcerated.

P. 901. Part of the 'ower end of a femur, from one of the condyles of which the whole of the articular cartilage has been separated in one piece. The surface of the cartilage next to the bone has many spicula separated from the femur attached to it; its free surface is healthy, and its thickness is not diminished. The surface of the bone from which it was separated is soft and very irregular; its subjacent cancellous tissue is filled with caseating inflammatory material.

P. 902. The upper end of a right tibia. Of the inner tuberosity a small portion has undergone necrosis; the dead part includes the proper articular surface, from which the cartilage has been removed, and is surrounded with a deep ulcerated trench. The whole of the articular cartilage has been removed from the inner facet, the bone being covered with connective tissue, through the medium of which it was probably ankylosed to the corresponding surface of the femur.

The outer facet is intact, as though this half of the joint had been isolated from the rest by adhesions.

P. 903. The lower end of a left femur with a small part of the surrounding synovial membrane. The synovial membrane covering the outer surface of the external condyle is greatly thickened; in parts it has a pulpy appearance, in parts it is opaque, friable and caseating. The soft tissues in the intercondylar notch have a similar pulpy appearance. The cartilages of the femoral condyles have been largely absorbed, but the compact bone beneath is unaltered. A layer of inflammatory exudate adheres to the patellar surface and the tissues in the intercondylar notch.

P. 904. A patella, from which one-half of the cartilage has been destroyed, leaving a flat surface having an appearance as if produced by friction. The exposed cancellous bone has been largely replaced by a soft friable substance, probably tuberculous granulation tissue, which is in places covered with a thin layer of inflammatory exudate. The rarefactive inflammation of the bone has resulted in a loosening of the cartilage from it at several points. The surrounding synovial membrane is little, if at all, thickened.

P. 905. The upper end of the right tibia of a child. There is extensive destruction of the internal articular cartilage, with deep ulceration of the subjacent bone. The exposed bone is very vascular and at one spot is an opaque white mass of, apparently, caseous material. The opposite articular cartilage is superficially eroded and pitted over a considerable area and in one or two places the bone has been exposed.

P. 906. The outer condyle of a femur. The outer surface and the groove for the popliteus muscle is covered with a layer of caseating granulation tissue. The articular cartilage at the edge of the layer is eroded and, at the anterior part, completely loosened. In the latter situation subchondral caries has extended over a large area of the condyle, with loosening of the cartilage. The free surface of the cartilage is also irregular from superficial erosion, and at its lower part is covered with fibrous adhesions.

P. 907. The patella and surrounding soft parts from the same joint as the preceding specimen. The mucous membrane is swollen and pulpy, and at the right border of the patella it has eroded the articular cartilage, which has become extensively loosened by subchondral caries, which is displayed by transverse incision and reflection of the cartilage.

"This person had a white swelling." *MS. Catalogue.*

P. 908. One of the condyles of a femur of a young subject. A large portion of its articular cartilage has been destroyed by subchondral caries. The remainder, thinned uneven on both its surfaces, and in several places perforated, is fixed to the bone by its outer border alone. The exposed surface of the bone is covered with granulations; its interior appears healthy. What remains of the synovial membrane is thickened and pulpy, and forms a deep swollen border overlapping the outer edge of the cartilage.

P. 909. One of the femoral condyles of a young subject, from which two large portions of the articular cartilage have been nearly detached by subchondral caries of the bone. In the middle of the condyle a large portion of the whole thickness of the cartilage has been completely destroyed, and its place occupied by a pulpy tissue, continuous with the thickened synovial membrane round the border of the articular surface. At the anterior part, also, the whole thickness of the outer border of one of the nearly detached portions of cartilage has been removed; and its place is in like manner occupied by the thickened synovial membrane, with which the soft tissue just mentioned is continuous.

P. 910. One of the condyles of a femur. At the margin of the cartilage bounding the intercondylar notch is a line of pulpy granulation tissue, probably tuberculous, which has eroded the edge of the cartilage. Anteriorly this has undermined the cartilage, a considerable portion of which has become loosened. The anterior and posterior parts of the articular cartilage are covered with fibrous adhesions.

Knee of Dog

P. 911. A vertical section of the knee-joint of a young Dog. The whole of the synovial membrane visible in the section is thickened by a somewhat nodular deposit, which is $\frac{1}{2}$ inch thick in places. In the upper part of the joint the thickened membrane has a villous form. The bones and articular cartilages are healthy, except at the posterior part of the femoral condyle, where there is a small deposit, over which the cartilage is depressed. The fibrous capsule of the joint is greatly thickened, producing externally the appearance of a tumour.

Microscopical examination of the thickened capsule shows circular areas scattered through the fibrous tissue, and composed of endothelial cells, in places separated by a delicate fibrous reticulum. In some of the nodules small giant cells are present and there is a narrow peripheral zone of lymphocytes. The lesion is a granuloma, but differs in some respects from typical human tubercle.

Ankle

P. 912. A vertical section of the lower end of a tibia. The articular surface, except for a small area at its centre, is covered with a thick layer of granulation tissue, which has apparently spread inwards from the capsule. The cartilage has been destroyed, but the compact layer of the bone is not perforated, nor the cancellous tissue invaded. At the upper part of the specimen is a small depression, the base of which appears to be formed by caseous material. At the back of the specimen a small caseous area is seen at the lower part of the section.

P. 913. A right astragalus. The upper and lateral articular surfaces have been extensively destroyed and the bone deeply ulcerated, and the main, central portions covered with fibrous adhesions. The anterior part of this surface, although denuded of cartilage, is not carious; in contact with it is part of the greatly thickened anterior capsule of the joint, which has been converted into a mass of caseating granulation tissue.

P. 914. Part of an astragalus, probably from a case of tuberculous disease of the ankle. The articular surfaces are surrounded by vascular granulation tissue and denuded of cartilage, except for two islands of cartilage on the superior surface. The compact tissue of the bone is not affected.

P. 915. A section of the same astragalus. The upper articular surface has necrosed for a depth of about $\frac{1}{4}$ inch and is demarcated from the living bone by a well marked groove. Necrosis has similarly affected the lower part of the articular surface of the head of the bone.

P. 916. A section of the preceding astragalus. There is slight caries of the upper articular surface.

P. 917. The remaining portion of the same astragalus, showing the necrotic areas mentioned in the last description.

P. 918. An os calcis, bisected in which several portions of the wall and of the superior articular surfaces have been destroyed by caries; the ulceration has penetrated so as to nearly isolate small pieces.

P. 919. A section of the preceding specimen, injected, dried and preserved in turpentine. A large area involving the upper and central parts of the bone is devoid of marrow and has become partly injected; this part has undergone necrosis and its spaces were probably filled with granulation tissue. Its trabeculae are continuous with those of the unnecrosed bone and there is no sign of separation at any part of its periphery. At one spot the trabeculae have entirely disappeared, probably through rarefactive osteitis.

P. 920. Section of an astragalus, probably from the same foot as the preceding, similarly diseased.

4. *Ankylosis*
P. 921-952

Knee

P. 921. A vertical section of a knee-joint, in which a small adhesion passes from between the condyles of the femur to the cartilage in front of the spine of the tibia. A shred of fibrous tissue is also attached to the anterior margin of the articular cartilage of the femur.

P. 922. A knee-joint, the cavity of which is completely obliterated by tough adhesions. The patella has been forcibly torn away from the outer condyle of the femur, to which it was united by bone. All the articular cartilages and the exposed surfaces of the bones (where they were not ossified together) are covered with an irregular layer of tough fibrous tissue.

P. 923. A vertical section of the knee-joint of a young subject. The articular cartilage of the tibia has been to a large extent destroyed and firm fibrous adhesions have developed, uniting the tibia to the intercondylar notch of the femur and the lower part of the patella. The capsule of the joint is thickened, especially at its hinder part, and its posterior surface has a somewhat lobulated appearance, due to unequal distension of the joint-cavity.

Ankle

P. 924. A vertical section of an astragalus and os calcis. A thin layer of fibrous tissue lies between the articular cartilages, which are themselves of natural thickness and appear healthy.

P. 925. A vertical section of the lower end of a tibia, an astragalus and part of an os calcis. Thin layers of fibrous tissue have formed in the articulations; the cartilages appear healthy.

P. 926. A portion of the lower end of a tibia, showing its articular cartilage of ordinary thickness and not altered in structure, but covered with a thin layer of fibrous tissue.

P. 927. An astragalus, with a narrow band of fibrous tissue passing from the outer part of its articular cartilage to the synovial membrane lining the posterior ligament of the ankle-joint. Similar small adhesions are, also, attached to the surface of the cartilage in front of this adhesion.

Bony Ankylosis—Spine

P. 928. Six dorsal vertebrae affected with osteoarthritis. The vertebral bodies are somewhat more porous than normal and their margins are thickened by a narrow deposit of new bone. They are firmly ankylosed together on the right side, apparently through ossification of the superficial ligamentous structures. On the left side the two lower vertebrae are similarly united. Most of the interarticular joints are affected and the articular processes are more porous than normal. The intervertebral spaces are reduced in width.

P. 929. Two cervical vertebrae affected with osteoarthritis. Ridges of bone project from the margins of their bodies and anyklosis has resulted from ossification of the anterior common ligament.

P. 930. A second and third cervical vertebra, the corresponding articular processes of which, on the left side, are united by bone, as the result of osteoarthritis. There is also some formation of new bone around the borders of the same processes on the right side, and the lower articular surface is enlarged and porous.

P. 931. Four dorsal vertebrae with parts of the ribs. Owing to ossification of the ligamentous structures, the costo-vertebral joints have become ankylosed; the vertebral bodies are similarly united on the right side. The ligamenta subflava are extensively ossified and the laminae of the lower two vertebrae ankylosed. The osseous substance of the vertebrae is light and porous from atrophy, and the spaces corresponding to the intervertebral discs are narrowed. Drawing No. 57.

P. 932. Four dorsal vertebrae, the bodies of which are united by a thick plate of bone extending over the middle and right side of their anterior surfaces, due, evidently, to ossification of part of the anterior common ligament.

P. 933. Five lower dorsal vertebrae affected with osteoarthritis. The osseous tissue of the bodies is abnormally porous and the body of the lowest vertebra is reduced in thickness. There is a lateral curvature, with reduction in the width of the left half of the vertebral bodies, and the bones have become ankylosed on the concave side of the curve, by means of stout outgrowths from the bones and ossification in the superficial ligamentous structures. The margins of the vertebral bodies show slight "lipping" in places and one of the lower inter-articular joints is markedly affected with the disease. Some of the articular facets for the ribs show atrophic changes.

P. 934. Two lumbar vertebrae of a White Bear, united by a broad plate of bone attached to the lower and lateral surfaces of their bodies and extending, like a bridge, across the intervertebral space. The bone has evidently been deposited in the ligamentous structures running between the surfaces of the two vertebrae.
Drawing No. 59, fig. 1.

P. 935. Two lumbar vertebrae of a Horse, united by a plate of bone attached to the adjacent surfaces of their bodies on the right side, and forming a considerable projection over the side of the intervertebral space. The new bone has evidently developed in the superficial ligamentous structures on the front of the spine. The facets of the upper articular processes are markedly diseased.

P. 936. Six dorsal vertebrae of a Horse. The bodies of five of them are united by new bone deposited in the anterior common ligament and forming a continuous thick and smooth plate investing nearly the whole surfaces of all the bodies and covering-in the intervertebral spaces, so that the outlines of the several vertebrae can hardly be discerned. On the sixth vertebra similar deposits of new bone are present but have not coalesced with those on the vertebrae above and below it. Similar extensive formations of new bone have taken place on the arches and between the spinous processes of some of the adjacent vertebrae, uniting them immovably together. The body of the highest vertebra is eroded, probably as the result of osteoarthritis.

P. 937. A vertical section of five dorsal vertebrae of a Horse, the bodies and some of the arches and spinous processes of which are united by stout plates of bone. The new bone, which appears to have been deposited in the common ligament, has coalesced with parts only of the surfaces of the bodies, and it does not in any degree encroach upon the intervertebral spaces or implicate the internal structure of the vertebrae. The facets for articulation with the ribs shows slight lipping due probably to osteoarthritis.

P. 938. The cervical and the first four dorsal vertebrae of a Lion, united by broad and thick plates of bone, irregularly formed upon the sides and lower surfaces of their bodies, and resulting, apparently, from ossification of the soft tissue covering the bones. Many similar plates pass from the body of one vertebra to the transverse process of another. The upper part of the spine is curved to the right side, apparently as a result of contraction of the fibrous tissue during the process of ossification.

P. 939. Three of the cervical vertebrae of an Ostrich, on which irregular plates and masses of bone have been formed and have united them immovably together. The bony deposits have occurred in the neighbourhood of the interarticular joints or intervertebral disc.

P. 940. Two cervical vertebrae of the same bird, similarly diseased and united.

P. 941. Two lower cervical vertebrae, from the same bird as the preceding specimen, showing similar changes. Drawing No. 58.

P. 942. A dorsal vertebra with the head of a rib, from the same bird as the preceding specimen. The rib was fractured and has become firmly united by bone to the vertebra.

Elbow

P. 943. A left humerus, radius and ulna. The radius and ulna are completely ankylosed by bone to the humerus, at about a right angle. The radius is semi-pronated; its shaft is flattened laterally and at its upper part is 1 inch in width, where it is continued on to the anterior surface of the lower end of the humerus without any differentiation into head, neck and tuberosity. At the junction of the two bones the lower end of the humerus is laterally flattened and presents a prominent crest continuous with the outer border of the radius. The upper extremity of the ulna is abnormally small and both the trochlear and capitular portions of the humerus are very defective. There is no evidence of disease of the joint and the deformity is probably developmental.

Hip

P. 944. Part of a right hip-bone, with the upper half of the femur. As a result of destructive disease of the hip-joint, complete bony ankylosis has occurred with the femur in a position of marked flexion and adduction. The section shows that the joint cavity is entirely obliterated by open cancellous tissue forming a continuous layer between the cancellous tissue of the neck of the femur and that of the floor of the acetabulum.

P. 945. The bones of a left hip-joint. There is complete bony ankylosis between the acetabulum and the head and neck of the femur, the latter being markedly shortened and widened. The femur is fixed in a position of flexion and external rotation. The section shows a continuous layer of cancellous tissue between the neck of the femur and the floor of the acetabulum.

Knee

P. 946. Parts of a femur and tibia, showing osseous ankylosis of the knee joint, due to destructive disease of the joint and upper extremity of the tibia. The latter is greatly enlarged by a deposit of new bone, which extends for several inches down the shaft. At the posterior part of the outer tuberosity of the tibia is the opening of a cloaca leading into the cancellous tissue. The bones are fixed at an angle of about 120°, and the tibia is displaced backwards slightly.

Ankle

P. 947. The left tibia and fibula, with the bones of the tarsus. There is slight angular deformity of the leg bones resulting from the old healed fractures a little below the middle of their shafts, and both bones are much thickened by a deposit of subperiosteal bone. The bones of the ankle and tarsal joints are firmly united by bone, and the os calcis shows evidence of disease in an irregular mass of new bone on its outer aspect.

P. 948. A right tibia, with the lower end of the fibula, the tarsal bones and the proximal ends of the outer four metatarsal bones. The foot-bones and lower extremities of the leg-bones are enlarged by chronic rarefactive inflammation, with extensive loss of substance in the lower end of the tibia and the three cuneiform bones. All the bones involved in disease are firmly ankylosed together by bone.

P. 949. The bones of a right ankle-joint, with the os calcis and scaphoid bones. Osseous ankylosis has occurred in all the joints as the result, probably, of disease of the body of the astragalus. The inner malleolus has been partly destroyed and the adjacent part of the astragalus is carious. The long axis of the os calcis is tilted, the upper part of the tuberosity being drawn up behind the extremity of the tibia.

Tarsus

P. 950. The bones of a tarsus, and three metatarsal bones, all united by bone, which both connects their substance and extends smoothly and continuously over their dorsal and plantar surfaces, and presents in some places a fasciculate appearance, as if the ligaments were ossified. The fifth metatarsal bone has been removed; and the corresponding articular surface of the cuboid bone is superficially ulcerated. The sole of the foot has lost its natural arch; but the several bones are not displaced from their mutual relations. Drawing No. 68.

P. 951. A cuneiform and metatarsal bone united by a column of bone in the position of the inferior ligament of the joint. The articular surfaces are healthy.

Toe

P. 952. The tarsal and metatarsal bones of a Horse, ankylosed by bone deposited between and around their articular surfaces.

Presented by Sir Philip G. M. Egerton (1806-1881).

5. *Osteoarthritis*
P. 953-961

Spine

P. 953. Four cervical vertebrae affected with osteoarthritis. Ridges of bone have been formed at the margins of the vertebral bodies and on some of the articular processes. The vertical measurement of the bodies, especially in the lower vertebrae, is reduced by atrophy and the osseous substance more porous than normal.

P. 954. A first lumbar and six lower dorsal vertebrae affected with osteoarthritis.

From their anterior borders, flat processes of bone have grown out and have met, without coalescing, in front of the intervertebral spaces. Continuous with these ridges new bone has been formed on the front of the bodies of the vertebrae. The osseous substance of the vertebral bodies is abnormally porous, but little deformity has resulted from the atrophy.

P. 955. Eight of the lower dorsal vertebrae of a Lion affected with osteoarthritis. New bone has been formed in ridges on the margins of the vertebral bodies and of several of the articular processes. The ossification appears to have occurred in the ligamentous structures.

Acromioclavicular Joint

P. 956. A left clavicle and scapula showing changes due to osteoarthritis. The joint surfaces of the acromioclavicular, sternoclavicular and shoulder-joint are slightly uneven, enlarged by a deposit of new bone at their margins, and perforated by numerous small apertures due to atrophy of the subjacent cancellous tissue. The spine and lower angle of the scapula are roughened by deposits of new bone.

P. 957. The right clavicle from the same case showing similar changes.

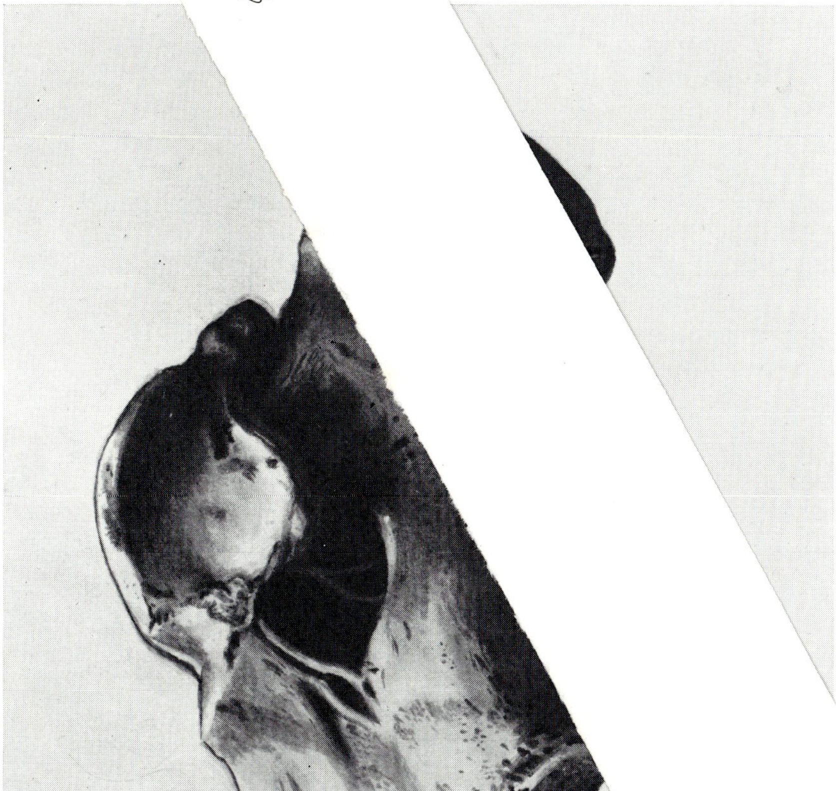

FIG. 13
"Anchylosis of all the Tarsal and Metatarsal bones; except the metatarsus of the great and little toes." Drawing No. 68 by William Bell. Specimen 950.

Elbow

P. 958. A radius and ulna showing the changes resulting from chronic osteoarthritis of the elbow and wrist joints. The extremities of the bones are much enlarged by the heaping-up of irregular coarsely nodulated masses of new bone round the margins of their articular surfaces. The surfaces themselves, as well as the new bone, are, in places, perforated by small worm-eaten apertures, and deeper excavations. The parts of the articular surfaces which are not ulcerated are smooth and hard, and have lost much of their natural form. In its lower third, the shaft of the ulna is uniformly enlarged and increased in density.

P. 959. The lower extremity of the right humerus from the same case as the preceding specimen. It shows changes similar to those in the latter with great deformity of the articular surfaces.

Knee

P. 960. The lower end of a femur with the patella from a case of traumatic osteoarthritis following partial outward dislocation of the tibia. The spine of the tibia, which has been lost, articulated with the deep groove in the external condyle. A considerable deposit of new bone has been formed round the articular surfaces and adjacent portions of the shaft, and the condyles are pitted and ulcerated apparently as a result of inflammation following the injury.

P. 961. "The patella of a man who died after having bruised his knee." *MS. Catalogue.*

The central portion of the articular cartilage is slightly swollen and irregularly fissured; the fissures pass through its substance in planes perpendicular to the surface of the bone. The change represents the early stage of chronic osteoarthritis, leading to fibrillation of the cartilage.

6. *Gout*
P. 962-969

Hand

P. 962. A carpus, with a slight gouty deposit on the articular cartilage of the trapezium and scaphoid bones. Some of the other cartilages are thinner than normal and slightly eroded at their margins.

P. 963. Parts of a carpus and metacarpus. In the thickened tissue over the bases of the metacarpal bones there are two irregular cavities, unconnected with the joints. These probably contained gouty deposits.

P. 964. A thumb, showing a gouty deposit in the cartilages of the metacarpo-phalangeal joint. The margins of the cartilages are eroded. Over the articulation of the phalanges is a cavity, which contained gouty matter and communicated with the joint.

P. 965. The proximal end of the first phalanx of a finger.

In the upper part of the articular cartilage, there is a narrow, semicircular, opaque zone due to the deposition of urate in its substance.

Knee

P. 966. A knee-joint, showing a gouty deposit covering nearly the whole of the articular cartilages of the femur and patella. A considerable quantity of the same material was visible in the recent state on the adjacent parts of the synovial membrane. The cartilages themselves are thin, and in some places fissured.

Ankle

P. 967. An astragalus, from a "gouty" person. The cartilages of the astragalo-calcaneal joint have a somewhat opaque and granular appearance but the characteristic white deposit is lost. The other cartilages show erosions of their central portions or margins.

P. 968. The os calcis of the same foot, similarly diseased. The opaque, granular appearance of the cartilages is better preserved in this specimen.

Foot

P. 969. Part of a tarsus, showing a gouty deposit in the proximal articular cartilages of the cuboid and scaphoid bones.

III

Muscles, Ligaments, Tendons, Bursae

P. 970-986

Muscle—Ossification

P. 970. The upper half of a femur longitudinally bisected.

From the anterior surface of the shaft, about 12·7 cm. (5 inches) below the summit of the great trochanter, there projects a low hemioval eminence of bone about 10 cm. (4 inches) in length. On one side the new formation shelves on to the general surface; on the other it overhangs it, and has evidently been produced in the crureus muscle. As shown in the section, the new formed bone is of coarsely cancellous tissue covered in with a thin compact layer.

The condition has probably arisen in a traumatic lesion of the crureus muscle. (R. J. Godlee, *Proc. Roy. Soc. Med.*, 1911, vol. iv, Surg. Sect., p. 144.)

P. 971. "A small spicula of bone from the diaphragm of a Horse." *MS. Catalogue.*

Note by William Clift: "It has not the appearance of having been an ossific deposit formed in that situation; but more probably has travelled from the cavity of the stomach, or some other viscus or part, similar to the progress which bones, needles, and other extraneous substances sometimes make."

Muscle—Secondary Carcinoma

P. 972. Portion of a pectoral muscle in which are several distinct, small, oval, hard and white tumours. The fasciculi of the muscle among which they are imbedded appear healthy.

The tumours are probably secondary deposits.

Patella Ligament—Ossification

P. 973. A patella, on the anterior surface of which new bone has been deposited, in the form of a broad thin plate which extends beyond its lower margin into the tendon.

P. 974. The patella of some mammal similarly, but more extensively, diseased. The new bone forms a thick mass, two inches long and half an inch wide, the ends of which are prolonged in numerous sharp parallel spines.

Tendon—Tenotomy

P. 975. A tendo Achillis which, together with the integuments behind it, appears to have been cut through. It has united by a tough, irregularly fibrous tissue, the surface of which is closely connected with the depressed and partially ulcerated cicatrix of the integuments.

Tendo Achillis of Deer: Experimental Tenotomy

P. 976. A longtudinal section through the tendo Achillis of a Deer which had been divided transversely by subcutaneous section. The ends are united by a band of dense white fibrous tissue of slightly lesser diameter than the tendon itself.

(This and the following four specimens may have been part of the experiment described in No. P. 109.)

P. 977. The other section of the same tendon.

P. 978. A longitudinal section of the tendo Achillis of a Deer which, it is believed, was divided transversely by subcutaneous section and in which the process of union is well advanced. The uniting medium is not distinguishable from the tendon itself, except by being less glistening, by its fibres being less regularly parallel and longitudinal, and by its surfaces being united with the surrounding fibrous tissue.

P. 979. A longitudinal section of the tendo Achillis of a Deer, which has been allowed to heal after subcutaneous section.

Tendo Achillis of Ass: Experimental Tenotomy

P. 980. Part of the tendo Achillis of an Ass, which was divided tranversely, but probably not by subcutaneous incision. The upper part alone of the tendon is shown: below it a tough and compact vascular substance, like that last described, extends nearly four inches downwards. One of the surfaces of this substance is covered with a thin layer of lymph, as if it had formed part of an open granulating wound.

Tendon—Gout

P. 981. "Gouty matter upon the tendon of a finger." *MS. Catalogue.*

P. 982. A similar specimen of a finger with gouty deposits on the tendon.

P. 983. A finger with gouty deposits on the extensor tendon near its attachment to the last phalanx, and a cavity which contained gouty matter over the articulation between the first and second phalanges.

Prepatella—Bursitis

P. 984. The anterior half of a bursa, from the front of a patella, enlarged and having its walls thickened, but smooth on their inner surface.

Tendon Sheath—Loose Bodies

P. 985. Several small, flat, hard bodies from a ganglion which was situated on a wrist.

P. 986. A large collection of "melon-seed" bodies from the sheath of a tendon.

IV

Mouth and Pharynx

P. 987-990

Tongue—Chronic Interstitial Glossitis

P. 987. A tongue with the fauces, larynx, and pharynx. In consequence of a chronic inflammatory fibrous overgrowth the surface of the tongue is thrown into convoluted folds and the thickened papillae stand out with unusual distinctness. At the root of the tongue the adenoid structures are enlarged from a similar cause, and extending up from those of the left side is a nodulated, tumour-like mass, part of which has been cut away to display its structure. The mass consists of a superficial zone of fibrous tissue, $\frac{1}{2}$ inch in thickness, which has destroyed the lingualis muscle and shades off into the deeper muscle, which is itself divided up by tracts and areas of fibrous tissue. The tonsils and adjacent mucous membrane of the fauces, the soft palate and the uvula are similarly thickened. The upper aperture of the larynx is reduced to a narrow vertical slit owing to a firm thickening of the mucous membrane of the posterior surface of the epiglottis, the ary-epiglottic folds and the covering of the arytenoid cartilages. The mucous membrane on the cricoid cartilage is wrinkled, apparently as the result of simple oedema. The larynx has been opened anteriorly showing that the thickening of the tissue is confined to the parts above the vocal chords, which are normal.

Microscopic examination shows that the tissue of the tongue is permeated with strands and areas of fibrous tissue in which the atrophic remains of the muscular fibres and other normal structures are embedded. The newly formed fibrous tissue exhibits considerable vascularity and is extensively infiltrated with leucocytes, of the type of lymphocytes and plasma cells.

Salivary Gland—"Mixed" Tumour

P. 988. Sections of a tumour removed from below the jaw of a young woman. It is of an irregularly oval form, with an uneven lobulated surface, and measures nearly two inches in its chief diameter. It is pale, compact, and firm, obscurely fibrous, and with several minute, smooth-walled cysts scattered in it. It is invested by a thin fibro-cellular capsule.

Microscopically the tumour is composed of epithelial tubular strands and irregular masses, widely separated in places by mucoid matrix resembling cartilage.

Lip—Squamous Carcinoma

P. 989. Part of a lower lip removed from a man at St. George's Hospital. A large portion of the border of the lip is covered with a slightly elevated warty growth. The border of the growth is sinuous, rises suddenly from the surrounding parts, is smoothly rounded, and a little everted. Except at the border, all the exposed surface of the growth appears excoriated or unevenly ulcerated.

Microscopic examination shows the tumour to be a squamous-celled carcinoma.

Tonsil—Hypertrophy

P. 990. A tongue, with the soft palate, fauces, and other adjacent organs. Both the tonsils are enlarged.

V

Oesophagus

P. 991-1000

Oesophagus—Congenital Stricture

P. 991. Part of a pharynx, with the larynx. Opposite the lower margin of the cricoid cartilage there is a projecting annular fold of the mucous membrane about $\frac{1}{10}$ inch in depth, narrowing the passage into the oesophagus. The adjacent parts appear healthy, and the fold is composed of healthy mucous membrane; the canal around the fold is slightly constricted.

P. 992. A pharynx and oesophagus, with the adjacent organs. Just below the cricoid cartilage the canal of the oesophagus gradually contracts to half its natural diameter, and then gradually dilates again. The mucous membrane at the contracted part forms transverse sharp-edged and projecting folds, which extend round the chief part of the circumference of the canal.

P. 993. A pharynx and oesophagus, with the adjacent organs. Below the cricoid cartilage the canal of the oesophagus is contracted to one-third of its usual diameter. The stricture is continued for about half an inch, and is abruptly terminated by a projecting transverse fold, below which the canal is uniformly dilated to much more than its usual size. At the stricture the mucous membrane appears somewhat condensed, and its surface is smooth and shining: above the stricture the walls of the canal are thickened by an increase of their muscular tissue; below it, in the dilated part, they are very thin, and their inner surface is creased and irregularly wrinkled.

Oesophagus—Carcinoma

P. 994. Parts of a pharynx and oesophagus, of which the canal just below the level of the cricoid cartilage is reduced to a diameter of a quarter of an inch, by thickening and contraction of the mucous membrane and the tissue beneath it. The thickening extends upwards to the level of the hyoid bone, and downwards in a less degree along the rest of the oesophagus. The interior of the canal, at the most contracted part, is ulcerated, rough, and flocculent: in two points, marked by portions of quill and whalebone, the ulceration has penetrated the walls of the canal, and has passed round to the front of the trachea, forming narrow sinuous fistulae. Baillie's *Morbid Anatomy*, fasc. iii, pl. iii, fig. 2.

P. 995. An oesophagus, of which the walls are increased to nearly half an inch in thickness by the formation of a layer of dense, greyish, semitransparent substance intersected with white fibrous transverse bands, between the muscular and mucous coats. The canal is contracted throughout its length to a diameter of one-third of an inch; the muscular walls appear healthy, the internal surface is rough but not ulcerated. The growth which exactly resembles the diffused cancer of the stomach, ceases at the lower end of the oesophagus, and the cardiac portion of the stomach is healthy. Baillie's *Morbid Anatomy*, fasc. iii, pl. iv, fig. 2.

P. 996. A pharynx, with the larynx and other adjacent parts. The posterior walls of the pharynx are occupied by a large, soft carcinoma, presenting a spongy texture, and deeply grooved and fissured on its surface. The walls of the upper and posterior part of the oesophagus are increased to about half an inch in thickness by similar disease. The anterior wall of the pharynx appears to have been very oedematous.

P. 997. A portion of an oesophagus, of which all the mucous membrane, through a length of about three inches, is destroyed by an ulcer of irregularly oval form, with slightly elevated sinuous margins, and a base which has a reticular aspect, and is intersected by prominent, coarse, fibrous bands. The middle of the ulcer is deeper than any other part of it, and here it has at one situation penetrated through the front of the oesophagus, making an oval aperture, to the margins of which a portion of the left lung is adherent. Beyond this aperture, also, the ulceration has extended for a short distance into the substance of the lung itself.

P. 998. An oesophagus, with the lower part of the trachea. The walls of the former, through a length of about four inches, are deeply ulcerated. The ulcer has the same general characters as that last described but its surface is softer, and is nearly covered with coarse flocculent shreds. It has penetrated through the front of the oesophagus, making in it an aperture an inch in diameter, to the margins of which the connective tissue of the posterior mediastinum is closely adherent.

P. 999. An oseophagus, with parts of the trachea, aorta, pulmonary artery, and right lung, all unnaturally adherent. An ulcer of cancerous appearance, commencing at the level of the bifurcation of the trachea, and extending about four inches downwards, has destroyed the left and a part of the posterior wall of the oesophagus. The remaining portions of the wall are rough but not flocculent, and all the rest of the canal is dilated and slightly thickened. There is a soft atheromatous deposit in the walls of the aorta.

P. 1000. Parts of a pharynx and oesophagus, with the adjacent organs. The whole of the mucous membrane, through an extent of about four inches below the arytenoid cartilages, is much thickened, and its internal surface is ulcerated, soft and pulpy. Lower down in the oesophagus there is a circular elevated ulcer, about half an inch in diameter, with a similar soft surface. The thyroid gland is enlarged.

VI

Peritoneum, Omentum and Mesentery

P. 1001-1012

Peritoneal Adhesion

P. 1001. A convolution of an ileum, partially laid open. At one part the opposite peritoneal surfaces are connected by a band of false membrane, round which a bristle is tied. There are several small circular ulcers of the mucous membrane.

Omentum—Cyst

P. 1002. A stomach, with the great omentum. The cavity of the omentum was distended by fluid. Its texture appears healthy and there is little fat on it.

Hydatid Cyst

P. 1003. A portion of omentum (not human), in which are two large calcified thick-walled cysts. The internal surface of the cyst, which is laid open, is wrinkled, rough, and of a yellowish-brown colour.

P. 1004. A portion of omentum, with a collapsed hydatid cyst between its layers.

P. 1005. A thin-walled, spherical membranous sac, containing an hydatid cyst, collapsed and beset on its inner surface with minute opaque white granules, which consist of groups of Echinococci. A portion of omentum is attached to the outer surface of the cyst.

Mesentery of Sheep—Hydatid Cyst

P. 1006. Several globular cysts, attached by long slender pedicles to the mesentery of a Sheep. They are formed of tough thin membrane; and one of them, which is laid open, is filled with the membranes of hydatids.

Omentum of Ox—Pedunculated Lipoma

P. 1007. A small, round fatty tumour, attached by a short narrow pedicle to the omentum of an Ox.

Peritoneum of Ox

P. 1008. A fatty tumour which was attached by a long narrow pedicle to the exterior of the intestine of an Ox. The tumour is composed of two chief parts united by a narrow isthmus, and each of an elongated oval form. It is about six inches in length, and at its widest part is two inches in diameter. The fat of which it is composed is firm, like beef-suet, and not lobulated.

P. 1009. The thin bilocular capsule of connective tissue in which the tumour last described was enclosed, together with the pedicle by which it was attached.

P. 1010. Section of a small, round, flattened tumour, attached by a small pedicle to the interior of the abdomen of an Ox.

P. 1011. The other section of the same tumour.

Peritoneum—Mucoid Carcinoma

P. 1012. A portion of great omentum which is thickened, so as to measure antero-posteriorly 1·8 cm ($\frac{3}{4}$ inch) by a diffuse carcinomatous growth the cells of which have undergone such extensive mucinoid degeneration that the persistent septa produce a coarse lung-like structure, the alveoli of which are filled with the translucent material. On the back of the preparation the intra-alveolar substance has been brushed out.

Histological examination shows the growth to be of alveolar construction, the meshes of which are filled with mucinoid material, about the margin of which, in certain places, groups of epithelial cells persist. The neoplasm is doubtless a carcinoma which was secondary to primary disease of one of the abdominal viscera.

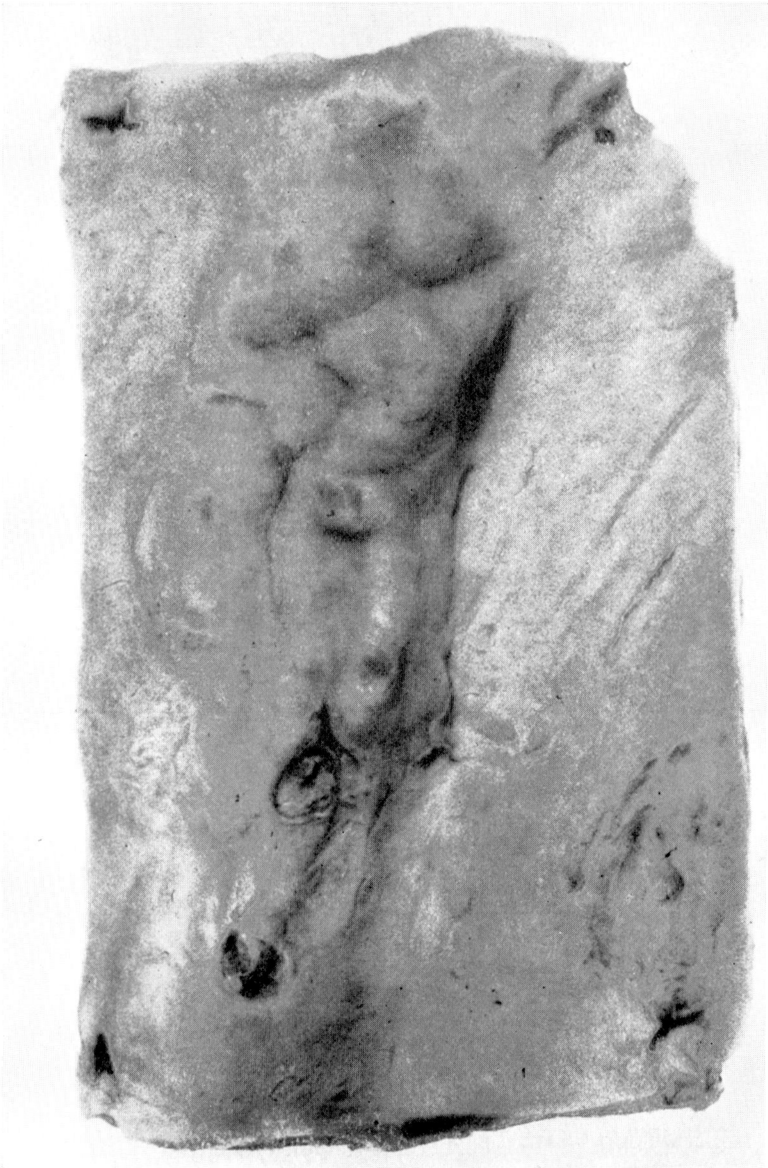

FIG. 14
Part of a human stomach, with a vein damaged by ulceration. Specimen 1015.

e p. 117

VII

Stomach

P. 1013-1031

Hair-ball

P. 1013. A spherical hair-ball $2\frac{3}{4}$ inches in diameter, probably from the stomach of a Cow. The regular disposition of the hair indicates that the mass has rotated during its formation, round one axis.

Arsenic Poisoning

P. 1014. Portion of the stomach of a young woman who died thirteen hours after taking half an ounce of arsenious acid. The appearance of the stomach shortly after death was thus described by Mr Hunter:

> "The stomach contained about twelve ounces of a greenish fluid, with a curdy substance in it. On the internal surface of the great curvature, near the cardia, a portion of the villous coat, about the size of a crown piece, was partially destroyed and of a dark red colour, with a regularly defined edge, and some of the arsenic adhering to different parts of its surface. The rest of the stomach was in a sound state."

Of these changes no trace is now visible, but the injured portion of the mucous membrane is less wrinkled than the rest, and some small shreds are adherent to it.

"The case of a young woman who poisoned herself in the first month of pregnancy, by Thos. Ogle, to which is added an account of the appearance after death, by the late J. Hunter," *Transactions of the Society for the Improvement of Medical and Chirurgical Knowledge*, vol. ii, p. 63 1800.

Gastro-oesophageal Varix

P. 1015. Portion of a stomach, in which an ulcer extended through the mucous membrane into a large vein, opening it in two places and giving rise to fatal haemorrhage. The vein has been injected and is large and varicose.

Ulceration

P. 1016. Part of the stomach of a Porcupine, in which are four ulcers, produced by the irritation of a large bezoar. Two of them are more than an inch in diameter, and the ulceration extending through the base of one of these has perforated all the coats. The bases of the other ulcers are smooth and thickly covered with exudate.

P. 1017. Portion of the pyloric part of the same stomach, exhibiting a small superficial ulcer, of which the base is covered by a greyish slough.

Chronic Peptic Ulcer

P. 1018. Portion of the stomach of a lady 22 years of age, showing an oval ulcer which has perforated.

The ulcer is sharply cut and somewhat funnel shaped, the aperture in the muscular coat being smaller than that in the mucosa, and that in the peritoneum the least of all; the latter has all the appearance of being a recent rupture due to mechanical causes occurring during life.

There is a thin layer of recent coagulated lymph over the peritoneum, for some distance beyond the perforation.

P. 1019. Portion of the stomach of a nobleman, in which, close to the pylorus, there is a small oval perforating ulcer, with smooth abrupt edges, through which the contents of the stomach had escaped into the peritoneal cavity. Near this ulcer, and directly over the pylorus, is a smaller ulcer of the same kind, which has destroyed only the mucous membrane. The adjacent tissues appear condensed and contracted towards the ulcer; the pylorus also is much contracted.

"This hole in the stomach must have been some time in forming, because its edges are thin, round and regular like those of a cicatrised wound. Perhaps it arose at first from an ulcer, and whilst the coats of the stomach in that part became thinner by suppuration (till they were quite eaten through), the edges or circumference of the sore were in a state of healing, and at last were completely cicatrised." *MS. Catalogue.*

P. 1020. Portion of the posterior wall of a stomach, in which there is a large irregularly bordered perforating ulcer. The outline of the ulcer is irregular, as if it were formed by the coalition of two or more oval ulcers. At its extreme circumference it has destroyed only the mucous membrane. Just within its outer edge, which is smoothly rounded, but abrupt, some of the muscular fibres are seen terminating abruptly at the margins of the deeper part of its base, which, to the edge of the perforation, is formed of the condensed peritoneal and subserous tissues. The edges of the perforation are thin and sharp. Behind the ulcer the stomach adheres closely to the pancreas.

P. 1021. Portion of a stomach, in the mucous membrane of which, below the lesser curvature, there are two nearly oval ulcers, each about half an inch in diameter. They have smooth sharp edges and obscurely granulated bases, on which, in one of them, the open orifice of a small blood-vessel appears. The adjacent mucous membrane is healthy.

Fibromyoma

P. 1022. A longitudinal section of the stomach of a Codfish, and of a large tumour attached to it. The tumour is of a somewhat oval form, and measures about nine inches and five inches in its two chief diameters: it appears to have grown between the coats of the stomach, for the outer coat may be traced for a short distance over its surface. The greater part of it is composed of a very dense and compact semi-transparent substance, traversed by fine bundles of white shining fibres, which are variously arranged but chiefly radiate from the part at which the tumour is attached to the mucous membrane of the stomach. The exterior of the tumour is formed of a thick layer of pale fibrous tissue enveloping that just described; and in its centre is a large irregular cavity, formed apparently, by ulceration or softening of its substance. The mucous membrane and the other adjacent parts of the stomach appear healthy.

Diffuse Carcinoma

P. 1023. The pylorus with the adjoining portion of a stomach and duodenum.

The mucous and submucous coats are considerably thickened and indurated by a diffuse carcinomatous growth which is limited on the distal side by the pyloric aperture, and thins away on the proximal. The circular fibres of the muscular wall are much hypertrophied in the region of the new growth.

Microscopic examination shows the neoplasm to be a spheroidal-celled carcinoma. A proper alveolar disposition is for the most part unrecognisable, the epithelial cells lying in narrow discontinuous lines in the thickened submucous connective tissue.

P. 1024. The pyloric end of a stomach much contracted and thickened. The thickening is chiefly in the submucous layer, and the overlying mucous membrane is smooth except in the pylorus where ulceration has commenced.

P. 1025. Part of a stomach and duodenum. The submucous tissue for four inches above the pylorus is much thickened by a dense fibrous formation to which the mucous membrane is closely adherent. A piece of omentum with some lymph glands is adherent to the outer surface.

P. 1026. The pyloric end of the stomach with part of the duodenum, showing similar diffuse carcinomatous infiltration. In addition, a flat rounded plaque of growth, with an ulcerated centre, projects into the lumen.

Carcinoma

P. 1027. The cardiac portion of a stomach, with part of the oesophagus. A soft ulcerated fungating mass of new growth surrounds and constricts the cardiac orifice. The oesophagus above the obstruction is dilated and hypertrophied.

P. 1028. The pyloric part of a stomach, with the first portion of the duodenum. The pylorus is surrounded by soft fungating mass of new growth which has ulcerated through the wall of the stomach. The ragged opening was originally closed by an adherent portion of the liver.

P. 1029. A woman's stomach laid open from behind. Growing from the anterior wall, close below the cardia, there is a large almost quadrilateral carcinoma with very prominent undulating borders and a sloughing base.

The lymphatic glands along the lesser curvature are considerably enlarged from secondary infection; and the liver is adherent to the stomach opposite the carcinomatous area. Histological examination of one of the enlarged lymphatic glands shows the neoplasm to be of the columnar-celled type.

Symptoms of gastric disease had existed for twenty-five years.

The jejunum from the same case is preserved in specimen P. 1033.

P. 1030. The pyloric portion of a stomach, with the commencement of the duodenum. Above the pylorus the cavity of the stomach is nearly filled with a soft lobulated mass of new growth, parts of which are deeply ulcerated.

P. 1031. Portion of a stomach showing a large, somewhat quadrilateral carcinomatous ulcer, the edges of which are raised, thickened, and sinuous. The wall has been so deeply destroyed at one spot as to be translucent.

Microscopic examination of the wall of the stomach immediately beyond the ulcer shows the neoplasm to be a columnar-celled carcinoma, for the most part atypical in character.

VIII

Small Intestine

P. 1032-1042

Ileo-caecal Valve—Oedema

P. 1032. A caecum, with a part of the ileum. The aperture of the ileo-caecal valve is large, and its margins are thickened as if oedematous. There is no other appearance of disease.

Jejunum—Acute Inflammation

P. 1033. Portion of jejunum, the coats of which are much thickened and appear consolidated. A considerable deposit, hanging in loose shreds, is attached to the surface of the mucous membrane.

About eighteen inches of the intestine were in the same state as this.

The patient died with the cancer of the stomach preserved in No. P. 1029.

Ileum—Acute Inflammation

P. 1034. Portion of ileum, in which there has been extensive superficial ulceration of the mucous membrane. The ulcerated parts may be distinguished in irregular strips and patches of a whiter colour than the intervening portions of the membrane, and smoother from the absence of villi. The ulceration has chiefly affected the membrane intermediate between the Peyer's patches and some of the latter, as well as some of the solitary follicles, are completely insulated by it. The intervening and adjacent tissues appear healthy.

P. 1034A. A portion of ileum, inverted, in which there is a very large and prominent patch of Peyer's glands. The follicles are well seen beneath the surface of the mucous membrane. The solitary glands, also, are large, and project far beyond the surrounding surface of the membrane. The villi are long and turgid.

This specimen probably exhibits only the healthy aspect of these parts during the active discharge of their several functions. It was at some time transferred to the Physiological Series where it was included in the section relating to the Organs of Digestion, with the number J. 1109.2.

Ileum—Perforating Ulcer

P. 1035. Portion of ileum, in which there is a small oval perforating ulcer. The aperture in the mucous membrane is much larger than that in the peritoneum; both are oval, and their long axes are placed at right angles to one another. The margin of the ulcer is in both coats definite and sharp; the base gradually shelves towards the aperture in the peritoneal coat. The adjacent tissues appear healthy.

Jejunum—Tuberculosis

P. 1036. Portion of jejunum, inverted, in which an ulcer (probably tuberculous), occupying the situation of a collection of Peyer's glands, has penetrated through all the coats. The ulcer is nearly circular, has a diameter of about half an inch, and its margins are smooth, rounded, and abrupt. The rest of the intestine appears healthy.

P. 1037. Portion of jejunum, with part of the mesentery. An oval ulcer, extending more than half round the intestine, has penetrated deeply through its mucous membrane, and has at one part perforated all its coats. The aperture, which is circular and about one-third of an inch in diameter, was closed with a thick layer of exudate, part of which is now reflected. Beneath the peritoneum surrounding the aperture are some small tubercles, and several lymphatic vessels, tortuous and distended with tuberculous matter, some of which may be traced in the same condition to three enlarged mesenteric glands.

Ileum

P. 1038. Portion of small intestine, with tuberculous ulcers of its mucous membrane, and some of its large lacteals full of opaque white tuberculous matter. The mesenteric glands were also affected.

Ileo-caecal—Tuberculosis

P. 1039. A caecum, with its appendix and the last portion of the ileum. The whole of the mucous membrane seems to be superficially ulcerated, for it is very thin and there is scarcely a trace of the villi. At several scattered points the ulceration has extended more deeply and has destroyed the whole thickness of the mucous membrane. This is especially the case at the ileo-caecal valve and in the appendix veriformis. The edges of the former are as if corroded; the cavity of the latter is at one part obliterated.

Colon—Tuberculosis

P. 1040. Portion of colon, from a person who is said to have recovered from dysentery, but the ulcers have more of the tuberculous than the dysenteric characters. Two of them, elongated and oval, have smooth margins and bases, as if nearly cicatrised. Beneath the peritoneum covering the larger of these two ulcers are some small tubercles. The other ulcers are small, round or oval, with level bases and elevated, smoothly rounded margins turned in over the borders of their bases.

Intestine of Sparrow—Tuberculosis

P. 1041. "Scrofulous tubercles on the intestine of a Sparrow." *MS. Catalogue.*
The specimen includes the proventriculus and gizzard which are similarly affected.

Jejunum—Carcinoma

P. 1042. Portion of jejunum, on the inner surface of which is a flat tumour, superficially lobed, with a sinuous border, occupying the whole circumference of the intestine, and about two inches wide. The tumour has a soft, obscurely fibrous structure, and part of its surface is ulcerated.

Microscopic examination shows the tumour to be a spheroidal-celled carcinoma. Below is suspended a portion of mesentery containing an enlarged lymphatic gland, from the surface of which hang a number of filaments.

Microscopic examination shows similar new growth in the gland. The presence of a plexus of large, thin-walled blood vessels suggests a haemolymph gland.

IX

Large Intestine

P. 1043-1055

Colitis

P. 1043. A caecum, with parts of the ascending colon and of the ileum. In the mucous membrane of the caecum and colon there are several small ulcers, of various shape, with sharp abrupt edges and smooth bases. The mucous membrane of the ileum is healthy.

Ulcerative Colitis

P. 1044. An inverted caecum, in which are several minute ulcers of the mucous membrane. They are of various shapes; some oval, some strip-like, some with irregular outlines. Some are quite superficial, but some extend through the whole thickness of the membrane, and have smooth, abrupt, not overhanging margins: their bases are level, and formed apparently by healthy tissue. The tissues adjacent to them also appear healthy.

P. 1045. A portion of the colon of a Lion, the blood-vessels of which have been injected. All its coats are thickened, and there are several small circular superficial ulcers of the mucous membrane.

P. 1046. A portion of the colon of a Lion. The greater part of the mucous membrane is destroyed by ulceration, and long shreds hang loosely from the exposed surface.

P. 1047. Two portions of colon, the coats of which are thickened and indurated. The surface of the mucous membrane is unusually wrinkled and granulated.

Colon—Bacillary Dysentery

P. 1048. Portion of colon, in which the mucous membrane is extensively ulcerated. The ulcers are for the most part isolated, and of round or oval form; they extend through the whole thickness of the mucous membrane, and have smooth bases and abrupt thick edges. They are very numerous, and in some situations several of them have coalesced, so as to insulate small patches and tracts of the swollen and granular-looking mucous membrane. The walls of the intestine are much thickened, and its canal is bent from its usual course by the adhesion of the peritoneal surfaces of two contiguous portions.

Colon—Diverticulitis

P. 1049. A portion of colon, of which the canal is at one part nearly closed by a thickening and contraction of its walls. At the seat of the stricture the mucous membrane is raised in thick folds, and several ulcers lead through it to narrow fistulous canals, which pass into the surrounding condensed tissues, in the course indicated by some bristles. Above the stricture are several small superficial ulcers of the mucous membrane.

Colon of Horse—Diverticula

P. 1050. Portion of the colon of a Horse, which has several small brown calculi imbedded in cavities in its mucous membrane.

Colon—Typhoid

P. 1051. Portion of colon, from a person who is said to have had dysentery. The ulcers are partially healed after separation of the sloughs.

The preparation is illustrated in Baillie's *Morbid Anatomy*, fasc. iv, pl. 2, fig. 4.

Ileo-caecal Junction—Typhoid

P. 1052. A caecum, with its appendix vermiformis and the last portion of the ileum, in each of which there are several small deep ulcers of the mucous membrane, probably typhoid ulcers nearly healed. Their bases are smooth, formed of the submucous tissue slightly thickened and granular, and the borders of their bases are covered by their elevated overhanging edges. The ulcers in the ileum have the form of small Peyer's patches; those in the caecum and its appendix are smaller, and either oval, round, or irregular in outline.

Colon—Carcinoma

P. 1053. Portion of colon, which is at one part abruptly contracted to such an extent that its canal is nearly closed. The contraction appears to have taken place around a small ring-shaped tumour of the mucous membrane. The tissues round the tumour are slightly thickened and indurated; above it, the canal of the intestine is considerably dilated; below it, contracted. The patient, a woman, died as a result of the stricture.

P. 1054. Portion of colon near its termination in the rectum. The canal, through a length of an inch and a half, is very nearly closed by the thickening and contraction of its walls round, a small flat and superficially ulcerated cancerous tumour. On the exterior of the contracted part some enlarged lymphatic glands are indicated by portions of bristles. The intestine above the stricture is dilated, and below it is contracted.

P. 1055. Portion of intestine (not human), of which the mucous membrane is elevated by a layer of small, closely set vesicles, filled with a pellucid fluid; an example, perhaps, of mucoid carcinoma.

X

Rectum

P. 1056-1071

Stricture

P. 1056. The lower part of a rectum, with the anus. On one margin of the anus are several large haemorrhoids, and the skin for a considerable distance round it is excoriated. At the right side of the anus is an appearance of a narrow granulating wound, as if a fistula had been there operated on. Immediately above the anus the canal of the rectum is suddenly and irregularly contracted to $\frac{1}{2}$ inch in diameter, but without any apparent change in the structure of its mucous membrane. Above the contraction it is unnaturally dilated, its coats are thickened, and the tissues around it appear rather indurated and confused.

Fistula

P. 1057. The lower part of a rectum, with the anus, at the margin of which are two large haemorrhoids. A portion of porcupine's quill is passed through a small, round, ulcerated aperture in the coats of the rectum into a cavity external to it, which opens by a large aperture in the skin at the margin of the anus. The coats of the rectum are thickened and consolidated with the surrounding tissues.

Recto-vaginal Fistula

P. 1058. A rectum, with a uterus, vagina, and other adjacent parts. The canal of the rectum is, through its whole length, contracted to about $\frac{1}{2}$ inch in diameter, and is surrounded by, and firmly united to, a large quantity of indurated cellular tissue and fat, by part of which it is unnaturally adherent to the posterior wall of the vagina. Nearly the whole of the mucous membrane of the rectum is superficially ulcerated, and its inner surface is smooth, hard, and without wrinkles. About an inch above the anus a circular aperture

K

leads straight from the rectum into the vagina. Around the margin of the anus the skin is raised in prominent folds, as if by haemorrhoids, and the orifices of the sebaceous glands are very large. At the orifice of the vagina there are several ulcerated passages beneath the mucous membrane adjacent to the fistulous communication with the rectum.

P. 1059. The female pelvic organs, inverted. About four inches from the anus a large ulcerated aperture through the walls of the rectum leads forwards by a narrow passage (marked by a piece of whalebone) into the vagina, and sideways by a wide canal into a large irregularly ulcerated cavity above and behind the broad ligament. The origin of this cavity is uncertain; its lining is ragged and all the tissues around it are hardened and confused. There is a prolapsus of the uterus and upper part of the vagina. The second of Houston's folds in the rectum is well shown.

Haemorrhoids

P. 1060. The lower end of a rectum, with the margin of the anus. A prominent transverse fold of the mucous membrane of the rectum, $1\frac{1}{2}$ inches in length, and $\frac{1}{2}$ inch above its anal margin, has been produced, probably by haemorrhoidal enlargement of the subjacent veins. All the dilated veins here cut across are filled with hard and compact clots of blood. The whole margin of the rectum is thrown into a fold like that above mentioned, but thinner and less prominent.

P. 1061. The lower part of a rectum, with large haemorrhoids around the margin of the anus. The haemorrhoids are nearly separated into two rows (the smaller ones lying above the larger) by a constriction extending round the rectum at the situation of the lower margin of the sphincter ani. The upper row are internal, the lower external piles.

P. 1062. The lower end of a rectum, exhibiting numerous minute circumscribed depressions, perhaps ulcers of its mucous membrane. Below this, and near the anal margin, the mucous membrane is raised in several prominent longitudinal folds by haemorrhoids; and below these, separated from them by a constriction, there is another row of smaller external haemorrhoids at the very margin of the anus.

P. 1063. Part of a rectum and an anus, the margin of which is surrounded with haemorrhoids.

P. 1064. The lower end of a rectum, with the anus, the margin of which is surrounded with haemorrhoids. Above the haemorrhoids the mucous membrane of the rectum is ulcerated and irregularly wrinkled; elsewhere it is healthy.

P. 1065. The lower part of a rectum, with the anus. The coats of the rectum are thickened and consolidated with the surrounding structures. Its cavity is contracted, and is further reduced by prominent folds of mucous membrane, which probably cover haemorrhoidal enlargement of its veins. Around half the margin of the anus are haemorrhoids covered with thin and slightly excoriated skin.

The patient had continual diarrhoea.

P. 1066. The lower part of a rectum, with haemorrhoidal tumours both within and beyond the margin of the anus. Part of the mucous membrane between and upon the haemorrhoids is ulcerated and ragged; two inches above the anus also is an oval aperture, with smooth thin margins, leading through the mucous membrane into a small cavity between it and the muscular coat. The cut edges of the preparation exhibit sections of clots of blood in the dilated parts of the haemorrhoidal veins.

Rectum—Carcinoma

P. 1067. Portion of rectum with a broad, slightly elevated, soft and spongy cancerous growth in its mucous membrane. The surface of the growth is ulcerated; its texture gradually merges into that of the surrounding healthy mucous membrane; the growth occupies nearly the whole circumference of the intestine.

P. 1068. A portion of a rectum with a cancerous growth in its walls. The surface of the growth is deeply and unevenly ulcerated, so that little remains except its elevated sinuous margin and obscurely fibrous base. The walls of the intestine are thickened; and on its exterior the tissues adjacent to the tumour are indurated and contracted.

Microscopic examination shows the tumour to be a columnar-celled carcinoma.

P. 1069. A portion of rectum, on the inner surface of which is a large shaggy cancerous ulcer. The central part of the ulcer has penetrated through the rectum deep into the tissues which are adherent and condensed around its posterior wall. The coats of the intestine are thickened and indurated.

P. 1070. Portion of a rectum with the urinary bladder and other adjacent parts. At the level of the prostate gland, and extending for nearly three inches upwards, is a deep irregular ulcer, which has destroyed the greater part of the walls of the rectum, and near its centre has penetrated the urinary bladder, forming an aperture through which a piece of bougie is passed. External to the rectum, the tissues adjacent to the ulcer are indurated and contracted. Around the aperture in the bladder, which is situated in the middle line just above the ureters, there is a flat soft tumour, nearly 2 inches in its greatest diameter, of an obscurely fibrous structure; and an inch higher on the wall of the bladder is another tumour of similar aspect, but of smaller size. With these exceptions the bladder appears healthy.

P. 1071. Part of the rectum of an Ox, with a section of a large cancerous tumour growing between its coats and projecting far into the cavity. The tumour appears to have been developed in the submucous tissue, for at its upper part a thin layer of mucous membrane is continued over its surface. The fasciculi of the circular fibres of the muscular coat are separated at the base of the tumour; they are also elongated, and many of them appear to be continued for some distance into its substance. The tissues external to the muscular coat are thickened and consolidated. The lower part of the tumour has sloughed and is deeply ulcerated.

XI

Hernia

P. 1072-1094

Femoral Hernia

P. 1072. A large hernial sac with a piece of skin adherent to the fundus. The lining is for the most part smooth, except in the neck and at the fundus where there is an irregular layer of inflammatory exudate.

P. 1073. "The internal orifice of a hernial sac, to show its valvular appearance"

MS. Catalogue.

The appearance is due to an annular thickening of the peritoneum nearly limited to the margin of the mouth of the sac, which is thus reduced to a much less diameter than any part of the sac.

Inguinal Hernia

P. 1074. A large inguinal hernial sac, with shreds of exudate attached to the upper part of its inner surface. At its lower part the sac divided into two pouches of equal size, of which one lies behind the other.

P. 1075. Part of a pelvis, with the right inguinal region, dried, and exhibiting the relations of a small oblique inguinal hernia containing intestine.

P. 1076. The sac of an inguinal hernia, with a portion of omentum adhering firmly to its neck by several thick adhesions. The omentum at the adherent part, but not elsewhere, is thickened and indurated.

P. 1077. A small hernial sac, with omentum adhering to its neck and inner surface. The contrast between the healthy omentum above the sac and that which is thickened and indurated within it is well shown.

P. 1078. A large inguinal hernial sac containing indurated omentum, which in several places is adherent to its walls by strong and well-formed bands.

P. 1079. The sac of an inguinal hernia filled with indurated omentum. Below the sac there is a small hydrocele of the tunica vaginalis.

P. 1080. The sac of a large congenital hernia. The exterior of the sac is uneven and sacculated, through the unequal yielding of parts of its walls: the testicle is situated at the lowest part.

P. 1081. The sac of a large congenital hernia, nearly full of healthy omentum, which is in several places adherent to its walls.

P. 1082. An inguinal hernia, with a hydrocele of the tunica vaginalis. The hernial sac, which is placed directly above that of the hydrocele, and projects a little into its upper wall, is opened from behind; it is filled with healthy omentum, which is adherent to its inner surface. The hydrocele is of small size, and the testicle is situated in the middle of the posterior wall of the sac.

P. 1083. The sac of a scrotal hernia, combined with hydrocele of the tunica vaginalis. The cavity of the tunica vaginalis has been laid open, and its anterior part turned back; in front, the tunica vaginalis is thickened; behind, it is very thin; a few shreds of exudate are shown, by which its surfaces were united; the testicle, reduced in size, is situated at the lower and posterior part. The hernial sac has descended to the level of the testicle, pushing down the tunica vaginalis before it, so that the sac of the tunica vaginalis nearly surrounds the sac of the hernia.

P. 1084. A very large inguinal hernia, from a Monkey. The sac contains a portion of colon with omentum, both in a healthy state. The testicle is directly below the sac.

P. 1085. Part of a hernial sac, from a Horse, containing a large portion of colon, which is adherent to what remains of the sac and has false membrane upon its surface.

P. 1086. A hernial sac, from a Bitch, containing small intestine, a large portion of omentum, and part of the uterus.

P. 1087. An inguinal hernia, from some large animal, with small intestine and omentum protruded into the cavity of the tunica vaginalis.

P. 1088. Part of the pelvis of a Lion, with a hernia of the fundus of the urinary bladder through the left inguinal canal. The hernia is exposed by the removal of the left wall of the abdomen.

Umbilical Hernia

P. 1089. Part of the anterior wall of a child's abdomen, with a small umbilical hernia.

Hernia—Strangulation

P. 1090. Portion of small intestine and mesentery, a part of which was strangulated in a hernia and retains the constricted appearance derived from the pressure of the neck of the sac. The surface is roughened by inflammatory exudate.

Baillie's *Morbid Anatomy*, fasc. iv, pl. vii, fig. 3.

P. 1091. A loop of small intestine which has been strangulated in a hernial sac. The tight constriction marks of the neck of the sac are retained, and shreds of inflammation exudate adhere to the peritoneal surface.

P. 1092. Portion of small intestine, part of which was strangulated in a hernia. Sloughing appears to have taken place on one of its surfaces, where its coats are very thin and perforated.

P. 1093. Portion of ileum, which was tightly constricted (probably in a hernia) and burst through a large oval aperture, nearly an inch above the seat of constriction. The intestine above the constriction is large and flaccid.

P. 1094. Portion of small intestine, in the coats of which is a large opening, with thick irregular margins, near the attachment of the mesentery. It is a rough ulcerated or torn aperture, such as might be produced in a hernia or an internal strangulation. Lymph is deposited on the peritoneal coat for some distance around the aperture.

XII

Intussusception

P. 1095-1097

P. 1095. Portion of the small intestine of a child, four years old, in which are three intussusceptions close to one another. In all of them the intussuscepted portion is directed downwards. There is no appearance of any change of structure, and they evidently occurred at or soon after death.

P. 1096. Portion of the small intestine of a Cat, in which are two intussusceptions within a short distance. In one the intussuscepted portion is carried upwards, in the other downwards. The coats of the intestine are healthy.

P. 1097. An intussusception of a large portion of ileum within the caecum and colon of a Dog. The walls of the intussuscepted portion are very thick and were turgid with blood. Its lower orifice is very small and is turned directly towards the attachment of the mesentery.

XIII

Liver and Gall-bladder

P. 1098-1123

Liver of Sheep—Ossification

P. 1098. Two flattened branching portions of bone from the liver of a Sheep. They were probably formed in connection with blood-vessels.

Note by William Clift: "Qu. whether they have been originally blood vessels or absorbents now become ossified? I once saw a beautiful example of the latter on the liver of a Horse, but Baker, the slaughterman, would not let me have it as he intended to take it to Professor Coleman; but in the succeeding night the rats destroyed it. Dr. Baillie told me he had never seen an instance of the kind."

Liver—Abscess

P. 1099. Portion of a liver and of the wall of a large circumscribed abscess in it. The wall is, in parts, fasciculated, and is lined with a thin and nearly smooth layer of firm granulation tissue closely attached to the substance of the liver.

Liver of Agouti—Streptothrix

P. 1100. Part of the liver of an Agouti, in which there are numerous small white areas. Histological examination shows the presence of branched filaments.

Liver—Congenital Syphilis

P. 1101. This specimen was included under the heading "Congenital Syphilis" in the 3rd edition of the Catalogue, all copies of which were destroyed when the College was bombed in May 1941.

Liver—Cirrhosis

P. 1102. Showing a finely nodulated surface. The nodular pattern is accentuated on the cut surface owing to escape of fat from the degenerate liver cells leaving the fibrous septa prominent.

"Part of the liver of a woman, the whole substance of which was of a lighter appearance than common, and is composed of small bodies of the size of millet seeds." *MS. Catalogue.*

P. 1103. Section of a liver, the substance of which is pale, indurated, contracted, and formed into round and oval nodules of various sizes, which project upon the external surface of the liver, and on the surface of its section appear separated by thin partitions of fibrous tissue.

Liver—Hydatid Cyst

P. 1104. Part of the liver of a woman, who died in St. George's Hospital, containing an enormous cyst which was filled with hydatids. The cyst measures, in its present collapsed state, nearly nine inches in diameter; it is formed of tough tissue, a line in thickness, and rough with lymph deposited upon its inner surface. Part of the liver is shown expanded upon the cyst, and the diaphragm is adherent to its surface.

P. 1105. Numerous hydatid cysts extracted from the previous specimen. The largest has been inverted to show clusters of minute cysts attached to the inner surface.

P. 1106. Several hydatids of various sizes, from a woman's liver.

P. 1107. Hydatids, from a human liver.

P. 1108. A single hydatid, from a human liver.

P. 1109. The liver of a Monkey, to the surface of which are attached several large transparent membranous cysts, each of which contains one or more hydatids.

P. 1110. Portion of the liver of a Cat, in the substance of which, and intimately united to it, is a thin opaque cyst containing a single acephalocyst hydatid.

P. 1111. Another portion of the same liver, containing a similar cyst and hydatid. The cyst is larger than in the preceding specimen, and projects on the surface of the liver, where a part of it was covered only with thickened peritoneum.

P. 1112. A large cyst, from the liver of a Lion, which contained numerous hydatids broken and rolled up. The walls of the cyst are of dense texture and their external surface is uneven and inflamed.

P. 1113. A portion of the liver of a Pig (?) in which there is a thin globular cyst containing a single hydatid.

P. 1114. Portion of the liver of some small animal, on the surface of which, and partly imbedded in its substance, there are several transparent membranous cysts containing hydatids.

P. 1115. Portion of the liver of some animal, in which are numerous very thin-walled cysts containing hydatids of various sizes.

For note on Hydatid Cysts see Part I, chapter XVI.

Liver—Secondary Carcinoma

P. 1116. Portion of a liver, in which are numerous round and irregular masses. The masses nearly fill the liver, and project upon its surface: many like them have been removed from smoothly circumscribed cavities in its substance, in which they were imbedded; some appear to be surrounded by distinct membranous capsules.

P. 1117. Another portion of the same liver.

Gall-bladder—Dilatation

P. 1118. A dilated gall-bladder, the mucous membrane of which has lost nearly all the naturally reticulated arrangement of its surface, and is in some places depressed, as if by the lodgment of calculi. The neck is also dilated.

Gall-bladder—Calculi

P. 1119. A gall-bladder, inverted. It had a large calculus in its fundus, through the influence of which its inner surface has lost its reticular structure and appears delicately fasciculated, as if by the development of bundles of muscular fibres beneath it. On some situations, fringes of pointed processes, like large villi and papillae, are raised from the mucous membrane.

P. 1120. A gall-bladder, with two large calculi distending it just above the cystic duct. In other parts its walls are thin, though not distended; its inner surface is not reticular, but fasciculated by muscular fibres strongly developed beneath it. The cystic duct is of natural size.

P. 1121. A gall-bladder, with adjacent portions of the liver, duodenum, and other parts. The gall-bladder is nearly filled by a large white nodulated calculus; its coats are thickened, contracted, and adherent to all the surrounding organs. A bristle is passed through the cystic and common biliary ducts. There are numerous small black spots, not clearly defined, in the liver, the mucous membrane of the duodenum, the pancreas, and other tissues.

P. 1122. A gall-bladder with its ducts. The cystic duct is distended by a calculus to a diameter equal to that of the gall-bladder, of which it thus appears as if it were a part. The fundus of the gall-bladder looks as if the calculus had been lodged in it for some time before it was forced into the cystic duct.

Gall-bladder—Biliary Fistula

P. 1123. Biliary ducts, together with a cyst containing a calculus two-thirds of an inch in diameter. The cyst is placed between the cystic and hepatic ducts, adherent to both but not communicating with either. Its walls are tough, uneven, and closely applied upon the calculus, which, it must be supposed, passed by ulceration from the gall-bladder, and had a cyst formed around it. All the bile-ducts are dilated and thin.

XIV

Lymph Glands and Spleen

P. 1124-1133

Lymph Gland—Chronic Inflammation

P. 1124. Subcutaneous lymphatic glands, which are described in the Hunterian MS. Catalogue as having gone through the processes of inflammation, suppuration, and ulceration, without being destroyed. They were taken from a Negro, and it will be observed that in the cicatrix over the diseased glands the black pigment of the epidermis has been reproduced.

Lymph Gland—Tuberculosis

P. 1125. Sections of three lymphatic glands, removed, together with the thickened and indurated tissue around them, from over the lower part of a man's parotid gland. The deepest of the three glands has undergone caseation.

Lymph Gland—Calcification

P. 1126. Portions of lymphatic glands adhering to the exterior of a large artery, and having undergone calcification.

Spleen—Calcification

P. 1127. A portion of the capsule of a spleen, dried, with a thin plate of calcareous substance in it.

Spleen—Rupture

P. 1128. "The spleen of a man who received a hurt, who became faintish and low, and died. Upon opening the belly there was found a great deal of extravasated blood, and on searching for the wounded vessel we found the spleen broke."

MS. Catalogue.

Spleen—Abscess

P. 1129. Portion of spleen, showing parts of the uneven ragged boundaries of two collections of pus in it. Its capsule is, in one part, nearly a quarter of an inch thick and very hard.

Spleen of Ox—Cyst

P. 1130. Portion of spleen (from an Ox?), on the surface of which is a small, thin-walled transparent cyst.

Spleen—Hydatid

P. 1131. An hydatid cyst from a spleen, inverted to show clusters of minute hydatids attached to the inner surface.

P. 1132. Acephalocyst hydatids, from the same spleen. One of them is inverted, and its inner surface is beset with minute white bodies, probably groups of echinococci. See P. 239 for case history.

Spleen—Tubercle

P. 1133. Portion of a spleen in which very minute, round, pale yellowish granules are thickly scattered.

XV

Thyroid Gland

P. 1134-1136

Thyroid—Nodular Goitre

P. 1134. An enlarged thyroid gland, with the adjacent organs. The whole of the gland is enlarged, but the right lobe much more than the left; it has extended round the trachea and oesophagus, passing backwards between the latter and the carotid arteries till its lobes very nearly meet. Its surface is uneven, marked by deep impressions, in which some of the muscles of the hyoid bone lay. The superior thyroid arteries are very large, and on the right side the inferior thyroid artery is twice as large as usual: the left inferior thyroid is not shown.

Thyroid—Carcinoma

P.1135. Section of a thyroid gland and of a larynx and trachea, with some of the adjacent parts. The thyroid gland is greatly enlarged by malignant disease. It surrounds the trachea, and it, or invaded lymphatic glands not distinguishable from it, extends downwards to the arch of the aorta and surrounds the innominate and right carotid arteries. The tumour has ulcerated into the trachea.

Thyroid—Teratoma

P. 1136. Sections of a large lobulated tumour, removed from the neck of a child. It contains numerous mutually compressed cysts.

XVI

Heart

P. 1137-1152

Heart—Bicuspid Aortic Valve

P. 1137. Part of an aorta, with the aortic valve. The valve is thickened; two of the flaps have fused at their contiguous surfaces and the resulting septum has undergone calcification.

Heart—Normal

P. 1137A. A human heart, dried after the injection of its cavities and great vessels with wax.

In the catalogue of the Pathological Series published in 1849 it is described as "A healthy human heart" (No. 3454) and as such it was transferred some time later to the Physiological Series with the number K. 190.1.

Heart—Fatty Infiltration

P. 1138. A portion of the left ventricle of a woman's heart. The fat of the auriculo-ventricular groove is increased in amount and a thick layer covers the whole surface of the ventricle. The wall of the ventricle is abnormally thin and in its lower half is infiltrated with fat, the muscular tissue at the apex being entirely replaced by it.

See *Dissections of Morbid Bodies*, No. 106. Clift Transcript, vol. III, p. 108.

Heart—Fibrinous Pericarditis

P. 1139. Section of the ventricle of a heart, on the surface of which there is a thin layer of finely reticulated fibrinous exudate.

Heart—Adherent Pericardium

P. 1140. Part of a heart, of which the pericardium covering the ventricle is covered with a thin layer of fibrous tissue, and in two places is adherent to the parietal pericardium by delicate fibrous membranes.

Heart—Bacterial Endocarditis

P. 1141. Portion of the base of a heart, with large masses of rather firm vegetations adhering to the mitral, and one of the aortic semilunar valves. The mass on the mitral valve extends on to one of the chordae tendinae, which has become severed by ulceration. The aortic valve is slightly thickened and the mitral valve contains several opaque areas, probably of calcification. There are also several small scar-like thickenings of the endocardium of the septum ventriculorum.

Heart—Sclerosing Endocarditis

P. 1142. Part of a heart, with the mitral and aortic valves. The cusps of the mitral valve are thickened and fused together in the form of a funnel in the centre of which is the stenosed mitral orifice, appearing as a narrow crescentic opening which would admit the tip of the little finger. The chordae tendinae are thickened and shortened, and the tips of the papillary muscles fibrotic. The left auricle is dilated and hypertrophied.

This specimen was possibly from a Miss Wood, details of the post mortem finding being given in the Clift Transcript, vol. III (Dissections of Morbid Bodies), No. 169, p. 186, as follows:

"The appearances on opening Miss Wood. She had been complaining for two years before death of an indisposition which was general, and increased. Was at times very ill, with an increase of these general symptoms. The particular complaints were, an interrupted pulse, oppression in breathing, and pain at the lower part of the sternum, attended with lowness and anxiety. When she got cold, or was otherwise indisposed, these complaints were more severe. She had towards the last a tension in the belly and died from an increase of all the above complaints.

"On opening the abdomen I found about a quart of serum and some flakes of coagulated lymph in it. The liver had some adhesions on its convex

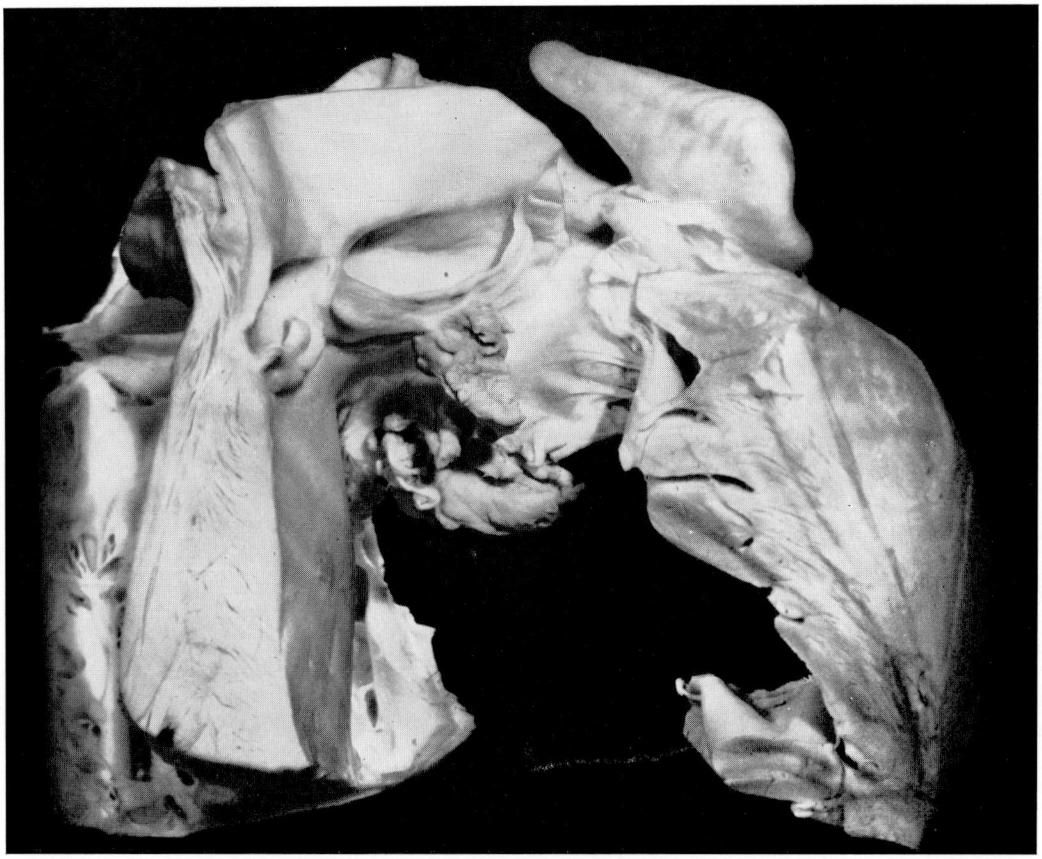

Fig. 15
The base of the heart showing the effects of bacterial endocarditis. Specimen 1141.

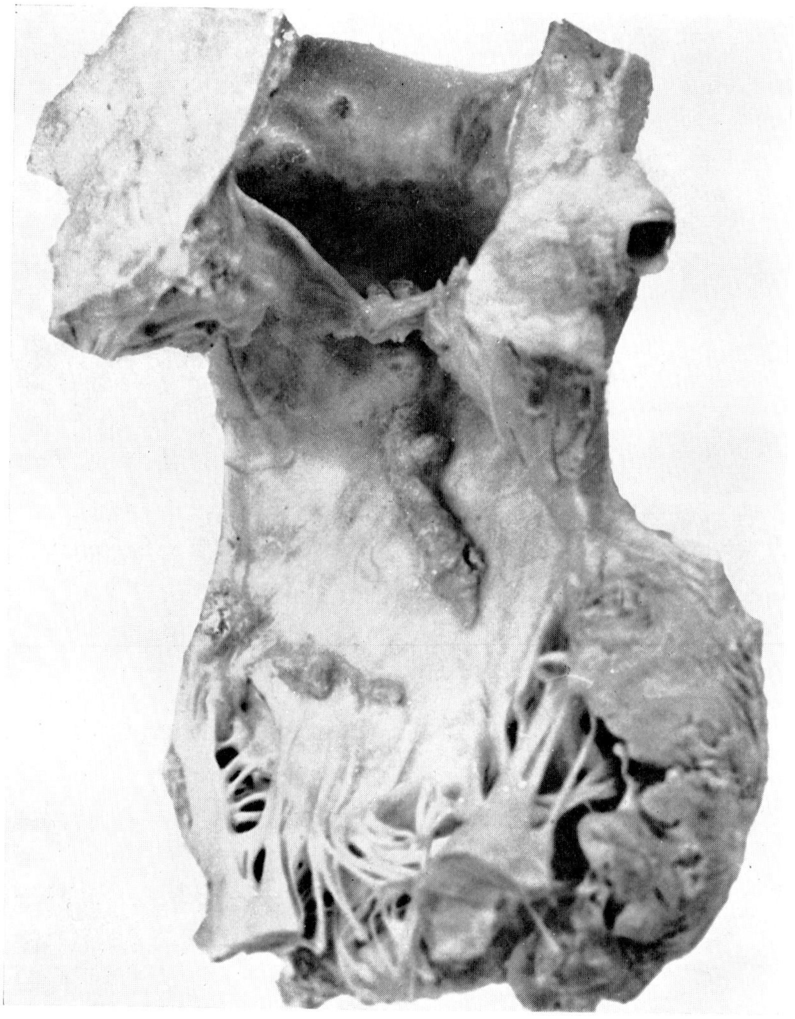

FIG. 16
Part of a human heart showing thickening of the mitral and aortic valves with deposits of calcareous matter. Specimen 1144.

surface and was rather larger than common. The spleen the same. The lungs adhered in many places to the pleura but otherwise pretty sound. There was about a pint of serum in the two sides (of the thorax).

"There was about two spoonfuls of serum in the pericardium. The heart was rather large; on its surface were many whitish spots and streaks made by the exudation of the coagulable lymph: a consequence of inflammation. The right auricle, ventricle and pulmonary artery were perfectly sound. The left auricle was thicker in its coats than common, particularly near the ventricle. The valvulae mitrales adhered to one another all round, were become extremely thick and hard, by which means they must have lost a considerable deal of their intended use: the blood must have found a passage back to the auricle. The beginning of the aorta was contracted, as also it- valves, which were also thickened and hard, and must have lost considers ably their use, by which means the blood in part would find a passage back into the ventricle.

"These two causes of regurgitation would produce considerable effects on the heart's motion, and were, most probably the cause of every other complaint and appearance.

"Dr. Heberden had been consulted and thought there were some symptoms of angina pectoris."

P. 1143. A portion of the left side of a heart with stenosis of the mitral valve. In the substance of the valve there is a mass of calcareous matter, with a rough ulcerated surface. The valve and the tendinous cords are thick, opaque, and rigid, and the orifice is contracted.

P. 1144. Part of a heart, including the mitral and aortic valves. Calcareous matter is deposited in coarsely granular masses and bars, which project upon the ventricular surface of the mitral and the arterial surface of the aortic valves. Both valves show slight general thickening.

P. 1145. The mitral and aortic valves, thickened, rigid, opaque and containing large masses of earthy matter. The aortic flaps are adherent to one another near their angles, and the aortic orifice is contracted to a circle of about four lines in diameter. The mitral orifice is also extremely stenosed.

P. 1146. Part of the left side of a heart, with the arch of the aorta. The transverse part of the arch exhibits several areas of atheroma; the inner surface of the ascending part of the arch is somewhat corrugated, probably as a result of syphilitic aortitis. There is considerable cylindrical dilatation of the ascending aorta; the aortic ring and orifice appear to be enlarged and the left posterior semilunar flap is laterally stretched; as a result of these changes the aortic valve was probably incompetent. The left ventricle is hypertrophied. The mitral valve is slightly thickened. The coronary arteries are atheromatous.

P. 1147. Part of a left ventricle with the aortic valve and first part of the aorta "from a servant of Lord Percy, who had palpitations of the heart, and died of a dropsy" (*MS. Catalogue*). The valves are uniformly thickened, and along their free margins is a line of minute vegetations, which have become organised. The surface of the interventricular septum below the valve exhibits several fibrous patches, slightly raised above the surface. Baillie's *Morbid Anatomy*, fasc. i, p.l 2, fig. 1.

P. 1148. An aortic valve, the semilunar flaps of which are thickened, especially at their free margins. In two of the flaps extensive calcification has occurred, rendering them rigid, and two have become adherent at their contiguous surfaces. As a result of these changes the aortic orifice is stenosed and reduced to a mere slit. On the ventricular surface of one of the flaps there is a small fibrinous deposit, the result of acute endocarditis.

Heart—Aneurysm

P. 1149. Part of a heart, with two aneurysms projecting from the upper part of the posterior wall of the left ventricle, near the septum, and just behind the mitral valve. The sac of the larger aneurysm was half-full of laminated coagula: its wall is formed only by the endocardium and cardiac pericardium, distended, thickened and adherent to the parietal pericardium; its cavity communicates with that of the ventricle by a circular aperture three fourths of an inch wide, with smoothly rounded margins. The aneurysm forms a hemispherical bulging towards the pericardium, measuring 2 inches in diameter. The smaller aneurysm lies immediately below the larger.

Heart—Coronary Atheroma

P. 1150. Part of a heart including the interventricular septum. The anterior descending coronary artery is large and its wall greatly thickened. Its branches are very prominent beneath the epicardium, and sinuous. The lumen of the main trunk of the vessel contains a clot, which has shrunk somewhat from the wall and is probably post mortem. The muscle of the left ventricle is extensively fibrosed, the cavity of the ventricle dilated and its endocardium thickened. There is a narrow line of atheroma at the commencement of the aorta. The cystic condition of the muscle is probably due to post mortem emphysema.

Heart of Bullock—Hydatid Cyst

P. 1151. Part of the left ventricle of a Bullock's heart, with a nearly globular cyst, two inches in diameter, embedded in it. The walls of the cyst are 3 mm. in thickness, pale, dense, and fibrous. Externally, it is intimately attached to the substance of the heart; internally, it is smooth, with some patches of calcareous matter on it. It contains portions of shrivelled membrane, part of which is opaque and brownish in colour, part white, delicate and transparent.

Heart of Pig—Cysticercosis

P. 1152. Part of the muscle of the ventricle of a Pig's heart, in which, embedded among its fibres, are numerous *Cysticerci*, in the form of oval cysts, measuring about ½ inch in the long diameter, and having a delicate transparent cyst-wall.

XVII

Arteries

P. 1153-1215

Artery—Traumatic Rupture

P. 1153. Part of a femoral artery, transversely divided, from which fatal haemorrhage occurred. The divided ends of the vessel are not more than 2 mm. apart and are wide open; the cavity of one portion of the artery has been exposed by a vertical incision, as if to show the unaltered state of the coats.

Artery—Secondary Haemorrhage

P. 1154. Parts of a femoral artery and vein, from a patient at St. George's Hospital who died of secondary haemorrhage after amputation. The artery has been divided through an elongated oval aperture in its wall, measuring half an inch in length; the edge of the aperture is smoothly rounded. The vessels are closely adherent and their walls thickened; the vein exhibiting transverse thickenings of the intima, the artery being atheromatous especially at its lower part. Towards the lower part of the vein a transverse fold projects into the lumen, and is probably a valve.

Artery—Experimental Division

P. 1155. Part of the femoral artery of a Boar. Beyond and adherent to the end, which was cut across during life, there is a hemispherical clot; but this is not, in the present state of the specimen, continued within the vessel.

"The crural artery of a Boar being divided, the bleeding ceased before the animal seemed weak; and upon examination this effect appeared to have been produced by a quantity of firm coagulated blood which had not only outwardly closed the orifice but within had regularly and conically contracted the diameter of the canal." *Parkinson's Lecture Notes*, vol. I, p. 10.

P. 1156. Part of the carotid artery of an Ass, which was divided in the middle of its course. The animal bled to death.

The divided ends of the artery have retracted for a distance of about 2 cm. ($\frac{3}{4}$ inch), and are contracted as a result of the muscularity of the vessel.

Umbilical Artery of Calf after Separation

P. 1157. The end of the umbilical artery of a Calf soon after its severance from the placenta.

The end which was distal in regard to the Calf, and which presents a jagged irregular line of division, is filled with coagulum which projects beyond the surface as a hemispherical eminence. The artery, which is 5 cm. (2 inches) in length, is still pervious, though diminished in size.

Artery of Dog—Experimental Ligature

P. 1158. Part of the carotid artery of a Dog, on which, probably some days before death, two ligatures were applied at the distance of 2·5 cm (1 inch) from each other.

The portion of the artery between the ligatures is contracted to about half its former calibre, and at the upper part is nearly filled with a conical clot, 1 cm. ($\frac{1}{3}$ inch) in length, which is closely adherent to the walls of the artery just below the upper of the ligatures. The ligatures had not separated when the animal was killed.

Artery—Effects of Ligature

P. 1159. Portions of an artery and vein, containing coagula. The coagulum in the artery has an elongated conical shape, as if formed above a ligature, but does not adhere to the walls. That in the vein completely fills the cavity, and is closely adherent to the walls.

P. 1160. Portion of a femoral artery and vein, from an amputation stump. The extremity of the artery is completely closed, and the tissues around it are condensed. A bristle has been passed beneath a short conical coagulum, the apex of which is adherent at one spot to the intima, whilst the base adheres to the vessel at its occluded end. The vessels have been coarsely injected; no injection has entered the clot, though some of it is adherent to its surface.

Parkinson's Lecture Notes, vol. II, p. 83.

P. 1161. Part of a femoral artery, which was tied at St. George's Hospital for the cure of a popliteal aneurysm. No obliteration or contraction of the artery has taken place. A small coagulum of blood, adherent to the lower end of the preserved portion, had probably formed above the ligature. The coats of the artery are thickened and the inner surface transversely grooved. The adjacent vein is filled with clot.

P. 1162. A femoral artery, with the adjacent parts, from a man whose leg was amputated at St. George's Hospital. Some days after the amputation haemorrhage from the artery occurred. A ligature, seen at the lower end of the specimen, was placed around that part of the stump from which the haemorrhage appeared to proceed, but the artery was not enclosed by it. The patient died after a few subsequent smaller bleedings and the end of the artery was found, as now shown, retracted $1\frac{3}{4}$ inches beneath the surface of the stump, and slightly contracted. For $\frac{1}{2}$ inch above the open extremity of the artery its interior is rough, and ulcerated; and above this, a conical clot is closely attached to its walls.

P. 1163. The end of a femoral artery from a stump.

"A boy at St. George's Hospital, about twelve years of age, had a white-swelling in his knee. He was becoming hectic. The limb was removed. He soon lost his appetite: no rest, sweats, &c.

"The stump often bled considerably, but when opened, it could not be perceived from whence it came. He became lower and lower, and at last died. On examining the part, I found the granulations ossified. I found that the artery had not adhered above the ligature, but had mortified above three-eighths of an inch, which was sloughing off. This artery must have been taken up pretty high to have been of any service."

Cases in Surgery, vol. i, p. 81.

Artery of Deer—Thrombosis

P. 1164. "Obliteration of the lower part of the iliacs, &c., of a Deer from the East Indies." (*MS. Catalogue.*) The upper ends of the internal iliacs are obstructed by a branched thrombus.

Artery—Stenosis

P. 1165. An irregular mass of bone, apparently of pathological origin, measuring $2\frac{1}{4}$ inches in its chief extent. On one side of it is part of a large artery, which is considerably narrowed at its site of adhesion to the bone. In this situation the arterial wall is atheromatous.

Aorta—Spontaneous Rupture

P. 1166. The arch of an aorta, with the roots of the large branches. In the posterior wall of the transverse part of the arch is a horizontal rupture about 1 inch in length, involving all three coats. For a short distance around the rent the wall of the vessel has been separated into two layers by the effused blood, as occurs in dissecting aneurysm. Localised thickenings are recognisable in the interior of the vessel, and the posterior end of the rent is continued as a narrow line of thinning of the wall, for a distance of 1 inch or more. The adventitia of the large branches and adjacent part of the aorta is thickened and indurated; the branches are thickened and their lining irregularly grooved by proliferation of the intima.

Artery—Medial Calcification

P. 1167. Part of a femoral artery the inner surface of which is raised into closely placed parallel ridges running transversely to the axis of the vessel, and caused by annular calcification of the middle coat.

Aorta—Atherosclerosis

P. 1168. The first part of an aorta, with the aortic valve. The aorta exhibits several small deposits of atheroma, some of which are calcified. There is also some thickening at the attached margins of the semilunar flaps.

P. 1169. The arch of an aorta, with the aortic valve. The distal part of the arch is extensively fibrosed and atheromatous, numerous calcareous plates being exposed within the fibrosed area. There is also a narrow line of calcification immediately above the aortic valve, and a broader zone of fibrosis with calcareous deposits somewhat higher up in the vessel. The semilunar flaps are slightly thickened and in one there is an irregular calcareous mass. One of the sinuses of Valsalva is dilated and the coronary artery arising from it is abnormally large and its walls rigid.

P. 1170. The arch of an aorta, with the trunk of the anterior coronary artery. There is a diffuse atheromatous change affecting the whole extent of the aorta. The whole of the intima is thickened and in its deeper layers are numerous atheromatous deposits, which in many places are ulcerated. The coronary artery is thick, rigid and tortuous. There is considerable dilatation of the sinus aortae and the innominate artery is abnormally wide and its walls thickened by atheroma.

P. 1171. Parts of the aortic arch and one of the carotid arteries. The inner surface of the aorta exhibits several irregular, slightly raised areas of fibrous appearance, in the deeper parts of which are calcareous deposits. At the bifurcation of the carotid artery there is an extensive atheromatous deposit the surface of which has broken down, producing an ulcer. The base of the latter is thin and the vessel in the situation of the ulcer is slightly dilated.

P. 1172. An aortic arch with part of the descending aorta extensively affected with atheroma. The inner surface is deformed by the presence of numerous, raised atheromatous plaques, which have coalesced so as to produce the appearance of a diffuse lesion. In the section of the wall the friable, yellowish atheromatous material can be seen lying in the deeper layer of the inner coat of the vessel. In several places the media has been separated to show that it is not affected. The orifices of the intercostal arteries are considerably narrowed by the thickening of the intima. The aorta itself is not sensibly altered in diameter.

P. 1173. A similar specimen, from a thoracic aorta. At the back of the specimen the wall of the artery exhibits localised bulgings in the situation of the larger atheromatous patches.

P. 1174. An abdominal aorta affected with atheroma. At the lower end of the vessel are several calcareous plates which have formed in the intima. Over most of these the endothelium has been destroyed.

P. 1175. A similar specimen, showing extensive atheromatous disease immediately above the bifurcation of the aorta. The superficial layer of the intima covering several patches of atheroma has broken through, producing large atheromatous ulcers. Much of the soft atheromatous material has been discharged, and the bases of the ulcers are thin and slightly bulged outwards.

P. 1176. Portion of an abdominal aorta showing the occurrence of extensive fatty and calcareous changes in the deeper part of the intima, which has undergone previous thickening from chronic endaortitis.

The inner surface of the vessel is raised in low elevations some of which are opaque and of a pale yellow colour from secondary degenerative changes. The fissures in the diseased coat at the lower part of the preparation have been produced after death.

In the divided edges of the vessel, the continuity of the media may be traced over the sites of disease. At the lower end on the right hand side, the outer, with the chief part of the middle coat has been reflected from the intima: in the dissection a thin lamina of the media has been isolated; this externally is smooth, and displays the normal transverse fibrillation, but to its inner aspect, a portion of the calcified material is adherent.

P. 1177. The lower part of an aorta with the common iliac arteries, extensively affected with atheroma. The inner coat of the aorta has been almost entirely converted into a rigid calcareous tube, which has become broken in many places, probably on opening the vessel. The iliac arteries exhibit a similar change, but in less degree.

P. 1178. The lower part of an aorta with the common iliac arteries extensively affected with atheroma. There is a soft friable deposit of atheromatous material involving diffusely almost the whole extent of the deeper parts of the intima. In several situations the deposit is a quarter of an inch or more in thickness, and in the iliac arteries causes considerable diminution of the lumen.

P. 1179. Part of an abdominal aorta, with the iliac, splenic and some other arteries (probably parts of those of a lower extremity), dried. There are numerous deposits of calcareous matter in the walls of all. Those in the aorta and common iliac arteries are in the form of scattered plates; but in the splenic and other smaller arteries the deposit takes the form of annular calcification, such as usually occurs in the muscular coat.

P. 1180. Thin plates of calcareous matter from the walls of an aorta.

Aortic Arch—Syphilitic Aneurysm

P. 1181. A section of the arch of an aorta, exhibiting a fusiform aneurysmal dilatation extending from one inch above the aortic valve to the commencement of the descending aorta. The dilatation commences and ends abruptly with well-defined borders, like those of the mouth of an aneurysm. The dilated part of the artery measures in its greatest diameter between two and three inches, and its length is about nine inches. The internal surface is uneven, and opaque, with deposits of atheromatous and calcareous matter. Two small secondary pouches have been dissected out on the posterior aspect of the specimen. In the distal part of the dilatation a large quantity of laminated clot has reduced the channel to the ordinary size of the aorta; some of the clot extends into the left subclavian artery. The orifice of the aorta and its valve are healthy.

P. 1182. The arch of an aorta, generally and considerably dilated from near its origin to the root of the left subclavian artery. Calcareous matter is thickly deposited in its coats.

P. 1183. The first part of an aortic arch with the aortic valve. The aorta is diffusely thickened and its inner surface rendered irregular by a fibrous deposit, probably of syphilitic origin. In the cut margin of the aorta the fibrosis is seen to extend into the middle coat. The orifices of the coronary arteries are reduced in size and immediately above one of the semilunar flaps is the opening of an aneurysm of the size of a pea. The aortic valve is slightly thickened.

P. 1184. Part of the front wall of a chest, with an aneurysm of the arch of the aorta, dried. The aneurysm arises from the front wall of the aorta, below the origins of the innominate artery, and the whole of the extra-pericardial part of the arch exhibits a fusiform dilatation. The aneurysm has produced absorption of the cartilages and ends of the first two left ribs, and of the left margin of the sternum, and has extended forwards beyond the wall of the chest.

P. 1185. The arch of an aorta, with parts of three left ribs, dried. The first portion of the arch exhibits slight fusiform dilatation and a saccular aneurysm, measuring $2\frac{1}{2}$ inch in diameter, projects from its anterior wall and protrudes slightly between two of the rib-cartilages.

P. 1186. Part of a chest, with a large aneurysm of the arch of the aorta. The whole length of the arch is dilated; it measures at one part $3\frac{1}{2}$ inches in diameter, and is nearly 7 inches in length; its coats are thickly beset with calcareous matter. The aneurysm arises from the most dilated part of the arch, from its upper and anterior wall, in front of the origins of the innominate artery. Making its way forwards and upwards, it has produced absorption of the upper third of the sternum, and of the cartilages and ends of the first three ribs on both sides. Within the chest the sac is 4 inches in diameter; it is constricted where it passes through the sternum; but beyond this it is enlarged to a diameter of 5½ inches and projects 3 inches beyond the walls of the chest. In its upper part there is a large aperture, through which it burst, after sloughing of the skin.

"*Case of Aneurism*

"Thomas Norman, a private soldier in the 8th Regiment, aged 55, says he had been a very healthy man. While on duty, at the latter end of the year 1782, he accidentally fell down while lifting a very heavy load, soon after which he felt a severe pain in his left side, which would frequently shift to the other, attended with difficulty of breathing, cough, and sickness at the stomach, and at times spitting some little blood. He had a sense of weight and uneasiness in the chest, about the middle of the sternum, for which complaint he had been frequently bled, which gave him some temporary relief. Medicines of various sorts had been given to him without the least benefit; vomits never relieved his sickness; blisters had been frequently applied, but to no purpose.

"About two years and a half after this accident, and about six months prior to his decease, he discovered a very small tumour on his sternum, nearly in its centre, and about the size of the tip of his finger, which rebounded very forcibly against it. He now made application to his officers, and from this time he did no more duty. On his admission to Chelsea Hospital, about three months before he died, the tumour on his breast had increased, from the upper part to the lower, for five inches, and from side to side much about the same. The tumour kept increasing in size from this time, so that a day or two before his death the circumference of the basis measured twenty-six inches; during the last four or five days the upper part of the clavicle began to slough, and continually discharged a small quantity of blood. The slough increased, and the discharge necessarily increased likewise, till he died. It may be necessary to remark that his urine for the last six weeks was so turbid as to take the appearance of a very strong decoction of bark.

"In most aneurisms the patients date them from some exertion they had been making, and, at the same time, feeling a pain in the part. However, it is most probable that that which they suppose a cause is rather an effect, and that the aneurism had made some advances, but was only felt at the time of the exertion."

Copied by Mr Clift from one of the lost Hunterian Manuscripts.

P. 1187. Part of a chest, with a large aneurysm of the arch of the aorta. The aneurysm rises from the front and right wall of the arch, just below the origin of the innominate artery, and the aorta adjacent to this part is generally dilated. The sac, the mouth of which is more than two inches in diameter, projects forwards and to the right side, nearly filling the upper part of this side of the chest, and making its way forwards by producing absorption of large portions of the first and second ribs, and of the side of the sternum. Portions of these bones, rough and ulcerated, project across the interior of the sac; above, it is in contact with the clavicle; in front, it projected several inches beyond the ribs; and there is an aperture in its posterior wall, through which, probably, it burst. The inner end of the right clavicle is displaced upwards by the sac.

P. 1188. A heart with the aortic arch, showing a large saccular aneurysm springing from the anterior wall of the ascending portion of the arch. The anterior part of the sac, which measures 4 inches in diameter, has been removed; its interior is lined with a layer of laminated clot. A row of bristles marks the aperture of communication with the aorta, which exhibits a marked general dilatation extending from its origin to that of the innominate artery, where it nearly regains its natural size. Inferiorly the sac adheres to the outer surface of the pericardium and posteriorly to the right lung; and the superior vena cava is stretched over its hinder wall. All the cavities of the heart are dilated and the right ventricle is covered with an excess of fat.

P. 1189. Part of a heart, with the aortic arch, trachea, roots of the lungs and part of the right pleura. Immediately above the aortic orifice, which is of normal size, the aorta exhibits a general dilatation involving the whole of the ascending aorta and measuring about $3\frac{1}{2}$ inches in diameter at its widest part. Distally it ends abruptly in a prominent transverse ridge. The dilatation affects chiefly the sinus of the aorta and the sac extends downwards and to the right, between the pulmonary artery and the right auricle, which is displaced backwards. The inner coat of the aorta in the affected part is thickened and fibrotic and in parts is covered with a thin layer of laminated clot. The superior vena cava is adherent to the upper part of the sac and the innominate left jugular and subclavian veins are obstructed by partly laminated, partly organised clot. The heart is covered with fibrous adhesions.

P. 1190. Part of a heart, with the first portion of the aortic arch. In the outer wall of the arch are two round openings, about 1½ inches in diameter, leading into saccular dilatations of the vessel. The anterior of the two pouches bulges slightly into the right ventricle and pulmonary artery, and a round perforation, ¼ inch in width, with smoothly healed edge, is present immediately above the margin of an adherent flap of the pulmonary valve. The aortic valve consists of two flaps only, both of which are thickened and deformed. The wall of the aneurysms contain numerous small calcareous plates and the intima of the aorta above the aneurysms is thickened and fibrotic. A small area of the wall of the anterior aneurysmal sac has been dissected, apparently in order to display a perforation.

P. 1191. The base of a heart, with the arch of the aorta, a section of the sternum, and other adjacent parts. About 1½ inches above the free margins of the aortic valves, the anterior wall of the aorta presents the orifice of a large aneurysmal sac. The orifice has an elongated oval form, and extends transversely across the wall of the artery; it measures nearly 1 inch in its chief diameter; the adjacent part of the artery is opaque white, but not otherwise changed. The aortic valves and the rest of the artery appear healthy. The aneurysmal sac is nearly spheroidal; from above downwards, and from side to side, it measures in each direction about 5 inches. It projects forwards and upwards; above and behind, it is in contact with the great arterial trunks, and its pressure has nearly obliterated the left subclavian; below, it is bounded by the pericardium; in front it is in contact with the sternum, which it has elevated to a distance of 3 inches from the aorta, besides making it unnaturally convex, and producing partial absorption of its substance. A great part of the sac contains firm laminated coagulum; and at the sternum its proper walls appear to have been completely destroyed. Everard Home: *Lecture Nores*, vol. ii, pp. 27-29.

P. 1192. Another section of the preceding aneurysm, exhibiting an irregular opening in the tissues above the first rib, through which the aneurysm burst.

P. 1193. The arch of an aorta, with the pulmonary artery, trachea. An inch and a half from the aortic valve, and just beyond the reflection of the pericardium the left and a part of the anterior and posterior walls of the aorta are dilated into a saccular aneurysm. The sac projects to the right upwards and backwards;

it has compressed and nearly closed the trunk and right branch of the pulmonary artery, the innominate artery, the vena cava superior and right auricle, and has pushed the trachea backwards and to the left side. The sac is partly filled with laminated clot. The parts around the aneurysm are indurated; the internal coat of the artery before and beyond the sac is thick, and wrinkled. The part of the aorta from which the sac proceeds is generally dilated in a slight degree, but immediately beyond it regains its ordinary size. The wall of the aorta between the aneurysm and the aortic valve, exhibits three small saccular depressions.

P. 1194. The arch of an aorta with the trachea. The posterior wall of the arch, from the origin of the innominate artery to the commencement of the thoracic aorta is dilated into a large, elongated, transversely oval, and flattened aneurysmal sac, which is partially filled with laminated clot. The sac compresses the trachea and has burst into it, just above the bifurcation, through a small smooth-edged oval opening.

Abdominal Aorta—Syphilitic Aneurysm

P. 1195. Part of an abdominal aorta, with a portion of the sac of an aneurysm. The aperture of the aneurysm is situated in the posterior wall of the vessel, immediately above the origin of the coeliac axis; it is oval in form and measures $1\frac{1}{4}$ inches in its longer diameter. The wall of the sac varies in thickness, in some places being very thin; at one spot there is a minute perforation. The lining has a crinkled appearance and at the lower part of the specimen is thickened, owing apparently to the partial organisation of an adherent layer of blood-clot. The wall of the aorta appears to be healthy. The diameter of the vessel below the aneurysm measures only four-fifths of an inch.

P. 1196. Part of the chest and spine, from the same case as the preceding specimen. The rest of the sac, preserved in this preparation, is of oval form, and measures 9 inches and 7 inches in its chief diameters. Projecting downwards, backwards and to the left, it has produced absorption of the last four dorsal and the first lumbar vertebrae. The eleventh rib has been displaced outwards and is embedded in the outer wall of the sac. The twelfth rib lies in the anterior wall of the sac at its lower part. The portions of aorta above and below the sac have been injected, and, like the portion from which the sac projected, appear quite healthy.

Aorta of Turtle—Aneurysm

P. 1197. The aorta of a Turtle, in which is a small conical aneurysm opening by a wide mouth into the aorta, and containing no blood-clot. The inner surface of the vessel above the aneurysm has well marked transverse ridges; at and below the aneurysm it is smooth and the wall of the vessel is thickened.

Laminated Thrombus from Aneurysm

P. 1198. Section of a large laminated clot partially decolorised, from an aneurysm of the aorta.

P. 1199. A large blood-clot from an aneurysm in a Lion. The laminae have been artificially separated.

Carotid Artery

P. 1200. A section of the upper part of a common carotid artery, with its bifurcation. A portion of the inner wall of the carotid about $\frac{1}{2}$ inch in length, near the bifurcation, is dilated into an oval aneurysm $1\frac{1}{2}$ inches in its chief diameter. The sac is nearly full of laminated clot. The inner surface of the vessels shows slight thickenings in places.

P. 1201. A large saccular aneurysm of the left external carotid artery, with the tongue, pharynx, larynx, and other adjacent parts. The aneurysmal sac is nearly globular in form, and measures $3\frac{1}{2}$ inches in diameter. It springs from the artery close to its origin by a narrow oval aperture, about one-third of an inch in length. A piece of whalebone has been passed from the common carotid artery, into the external carotid. The wall of the sac is a quarter of an inch thick and formed of dense fibrous tissue. Its cavity is nearly full of firm laminated coagulum. The sac has displaced the larynx, the upper part of which is markedly oedematous.

Subclavian Artery

P. 1202. A right subclavian artery, with a small portion of the sac of a large aneurysm which communicated with the artery near the origin of the thyroid axis. The trunk of the artery, which is laid open, shows several patches of atheroma. The sac is lined with laminated clot.

External Iliac Artery

P. 1203. An external iliac artery, with part of an aneurysmal sac, apparently of fusiform shape. The coats of the vessels above the aneurysm are thickened, extensively atheromatous and in places calcified.

Femoral Artery

P. 1204. Parts of a femoral artery and vein, injected and dried, together with a portion of a large bilobed aneurysmal sac, which was situated in the upper part of the thigh.

Popliteal Artery

P. 1205. The sac of a popliteal aneurism, with the arteries above and below it injected and dried. About half the circumference of the artery, and three inches of its length, are dilated into the sac. The vein, in close contact with the sac, is obliterated.

P. 1206. A popliteal artery, which is dilated in its whole circumference into a large aneurysmal sac. The cavity of the artery is exposed above the sac, and appears healthy, but its opening into the sac (marked by a bristle) is very small. The lower opening into the sac is about 3 inches below the upper, and beyond the sac the artery is almost obliterated. The sac has an oval form, and is more than 6 inches in its longer diameter; it is half full of laminated coagula.

P. 1207. A section of a large mass of clot from an aneurysmal sac, possibly from that last described. The main part of the clot is laminated, but at the upper part it contains a centre of more homogeneous and more recent clot.

P. 1208. Another section of the same clot.

P. 1209. Parts of a popliteal artery and of the sac of a popliteal aneurysm. The artery above the sac is larger than is natural, and thickened and fibrosed. At the margin of the sac the inner and, probably, also the middle-coat of the artery presents an abrupt well-defined border, as if they had been destroyed by ulceration. Below the sac the artery is nearly obliterated: a bristle is passed through its canal.

P. 1210. Part of a bilobed aneurysm, measuring nearly 3 inches in its chief diameter. The sac is nearly filled with laminated clot.

"Coagulum of an Aneurysm operated upon at St. George's Hospital with success." *MS. Catalogue.*

There is no statement as to the nature of the operation, or of the subsequent duration of life.

P. 1211. A section of coagulum from the sac of the same aneurysm.

P. 1212. Part of a popliteal artery. A bristle has been passed through an orifice by which the artery is said to have communicated with an aneurysmal sac. The inner coat of the artery is transversely ridged probably as the result of annular calcification. The popliteal vein opposite the remains of the sac is thickened and its lumen reduced.

P. 1213. The adjacent halves of the femur and leg bones, with some of the soft parts including a popliteal aneurysm, dried. The sac is irregularly globular in form and measures 3 inches in diameter. The popliteal and other veins are dilated, and adhere to the posterior surface of the sac.

Operation for Popliteal Aneurysm

P. 1214. "Femoral artery, operated upon for popliteal aneurism" (*MS. Catalogue*). Its coats are thickened and rigid. The ligature was probably applied to the middle of that part of the artery which is laid open. There is a conical clot of blood both above and below the line of ligature.

Parkinson's Lecture Notes, vol. ii, p. 83; *Pathology Catalogue*, Part 1, (1966), Chap. XXIII.

P. 1215. Part of a popliteal artery, from the same case as the previous specimen, but from the opposite limb. The adventitia is thickened and indurated. The inner coat shows raised patches of fibrous appearance beneath which yellowish deposits of atheroma can be seen in the section of the wall of the vessel. At the upper part of the specimen the inner surface of the artery is marked by transverse lines, probably resulting from annular calcification of the media.

This preparation is intended to show the changes preceding the formation of an aneurysm.

P. 1215A. A finely branched clot of blood from the vessels of a Horse.

This specimen was originally in the Pathological Series and was transferred in about 1900 to the Physiological Series with the number K. 12.1.

XVIII

Veins

P. 1216-1225

Varicose Veins

P. 1216. The veins of a leg, dilated, elongated and very tortuous. Their coats are thickened. One of the veins, which is laid open, has a thin layer of blood-clot deposited upon part of its inner surface, and at its upper part a dilated portion is filled with a globular clot of blood. The valves in this vein are shrivelled and contracted. The subcutaneous tissue adjacent to the small piece of skin preserved in the specimen appears to have been the seat of an abscess.

Varicocele

P. 1217. The injected veins of a testicle and spermatic cord which are greatly dilated and tortuous and form inferiorly a convoluted mass or varicocele, measuring $2\frac{1}{4}$ inches in its chief diameters.

Vein of Ass—Aneurysm following Venesection

P. 1218. A dried portion of the jugular vein of an Ass, from which blood had several times been drawn. On one side, probably that through which the punctures were made, its coats are distended into several round partial dilatations, like small aneurysms.

P. 1219. A dried portion of the jugular vein of an Ass, from which blood had been drawn several times.

There is a localised saccular dilatation, probably situated on the punctured aspect of the vessel.

Vein—Chronic Phlebitis

P. 1220. The veins of a leg. They are rigid and somewhat tortuous.

The wall, as seen in a transverse section of one of the branches, is greatly thickened and the lumen markedly reduced and obstructed by thrombosis, one of the lower branches which is opened longitudinally is partly filled with a thrombus.

Vein—Thrombosis

P. 1221. Part of a vein, the cavity of which is in one situation filled with a thrombus an inch long, with small portions of calcareous matter embedded in it. In other situations blood-clot is deposited in a thin uneven layer upon the inner surface of the vein. The coats of the vein are thickened and wrinkled.

Vein—Phlebolith

P. 1222. Part of a small vein, opened to show a partially calcified thrombus.

P. 1223. Five round phleboliths, "from peritoneal veins, human". One of them is elongated, and a small conical portion of dried blood is attached to its extremity, indicating that the calcareous matter of which it is formed was deposited within a thrombus.

Vein—Invasion by Tumour

P. 1224. Portion of a tumour from the abdomen of a Horse, with part of a large adjacent vein. The cavity of the vein is filled by a thrombus, the external part of which is in concentric layers. At the back of the specimen is a smaller vein, seen in transverse section, filled partly with blood-clot, partly with white new growth.

Microscopic examination of the thrombus shows it to contain masses of tumour cells, in places containing pigment.

P. 1225. Part of the arch of an aorta, with its large branches, and of the left internal jugular vein and pneumogastric nerve. The left common carotid and internal jugular, with the pneumogastric nerve, are embedded in a mass of malignant new growth. The vein has been opened; the lower part of the lumen exposed is occupied by growth and its upper part contains small thrombi.

Microscopic examination shows the growth to be a spheroidal-celled carcinoma.

XIX

Nose

P. 1226-1236

Nose—Syphilitic Ulceration

P. 1226. Portion of a skull, with the nasal mucosa in situ. The greater part of the septum nasi has been destroyed by syphilitic disease; the edge of what remains of the soft structures is in process of healing, and beyond this there projects a piece of the perpendicular plate of the ethmoid, which is bare and necrosed.

The greater part of the hard palate is likewise bare, dead, and in process of separation. On the buccal aspect of the palate there is a circular perforation in the soft structures, which is overlaid by the sequestrum referred to.

There is the scar of an ulcer in the mucosa over the inferior turbinal.

P. 1227. The anterior part of a skull exhibiting extensive destruction resulting from syphilis.

The bones which are lost comprise the greater part of the ethmoid, the vomer, and triangular cartilage, the nasals, the nasal processes of the superior maxillae, with the inner walls of the antra, the inferior turbinals, and portions of the hard palate, which is perforated anteriorly on either side.

The mucous membrane around each perforation is smoothly healed. There are two narrow persistent strips of mucous membrane at the site of the posterior border of the septum.

P. 1228. A septum nasi, in the cartilage of which there is an oval aperture 1·5 cm. ($\frac{1}{2}$ inch) in chief diameter, the result, it was considered, of syphilitic ulceration. The edges of the defect are smoothly healed.

Nose of Horse—Glanders

P. 1229. Part of the septum nasi of a Horse, showing extensive ulceration due to Glanders. At the sites of the larger ulcers, the septal cartilage is exposed, and in places eroded by granulation tissue.

P. 1230. One of the turbinals of a glandered Horse.

The mucosa on the convex aspect is widely ulcerated, the ulcers having shreddy necrotic bases, and being of very irregular ill-defined form.

P. 1231. Part of the septum nasi of a glandered Horse, the mucous membrane being stripped off from one of its aspects.

In the mucosa there are shown the scars of two or three extensive ulcers, around which the membrane is thrown into radiating folds as a result of the contraction attending their healing. The central part of one of the higher areas is still unhealed.

Nose—Mucous Polyp

P. 1232. Parts of the facial bones of the right side including the upper part of the antrum. A small soft gelatinous polypus is attached by a pedicle to the mucous membrane lining the outer wall of the antrum.

P. 1233. Mucous polypi, from a nose. Three or four separate polypi, connected with each other at their attachment to the mucous membrane, appear to have been extracted at the same time without breaking their connections.

P. 1234. A rounded mucous polypus, which was extracted from the nasal fossa.

P. 1235. An ovoidal nasal polypus measuring $1\frac{1}{2}$ inches by 1 inch. The greater part of it consists of a single cyst, to the upper part of which is attached a crescentic, solid portion of the tumour.

Nose—Chondrosarcoma

P. 1236. A sagittal section of the fore part of a boy's head, displaying the right nasal fossae. A polypus was removed from the hinder part of the fossa, where it has produced a marked depression of the outer wall. In front of this a smaller polypus is still attached to the upper wall of a deep depression. The inferior turbinal has been displaced upwards and flattened by the larger polypus, and it and the upper turbinals are involved in a cartilaginous tumour which has apparently arisen in the outer wall of the nasal fossa, and ethmoidal cells. The tumour forms an elongated mass, $2\frac{1}{2}$ inches in length adhering to the roof of the fossa; its anterior half is about ½ inch in thickness and ends anteriorly in a smooth rounded extremity projecting freely into the fossa; its posterior half is circular in section and one inch in diameter, and it adheres posteriorly to the body of the sphenoid bone, the anterior wall of which it has destroyed. The tumour has not penetrated to the cranial cavity. At its upper part it has undergone slight cystic degeneration. Hidden by the anterior extremity of the tumour is a small mucous polypus which is attached to the base of the middle turbinal.

Microscopic examination of the tumour shows it to consist of masses of hyaline cartilage of various sizes, separated by a rather loose connective tissue. which changes in places to myxomatous tissue.

XX

Larynx

P. 1237-1246

Larynx—Wound

P. 1237. The tongue, larynx, pharynx, and a portion of the integuments from a man who cut his throat between the hyoid bone and the thyroid cartilage. A large aperture remains in the situation of the wound, exposing the anterior surface of the epiglottis, which is uninjured. The wound is completely healed and the surrounding skin has contracted inwards, so that the lower border of the cicatrix is smoothly continuous with the anterior surface of the epiglottis. The mucous membrane of the upper aperture of the larynx is markedly oedematous.

Larynx—Ossification

P. 1238. A larynx with the upper end of the trachea, dried. The whole of the cartilages, with the exception of two symmetrical oval areas near the angle of the thyroid, have undergone complete ossification ensuing upon senile calcification.

P. 1239. The lower part of the preceding trachea with the right and left bronchi. Nearly all the cartilages have undergone similar changes.

P. 1240. A thyroid cartilage, partially ossified.

P. 1241. A thyroid cartilage, completely ossified.

Larynx—Perichondritis

P. 1242. "A piece of bone coughed up by a nobleman, Lord Polworth."*

"Under what circumstances is not recorded, but without further explanation, the impression would naturally be that it was a bony concretion formed there and expectorated, but this does not appear to have been the case. This substance is very dense bone; one part of it has a very high degree of polish, and has more the appearance of a portion of a decayed tooth than that of any other kind of bone. It must be useless to conjecture, were it of any import, or it might be conceived that a portion of a tooth that had become detached during sleep, and having passed into the fauces or trachea, on producing irritation was expectorated." *MS. Catalogue.*

The specimen is pyramidal in form, the base consisting of a smooth, concave articular surface, and is evidently a necrosed arytenoid cartilage.

* Baron Polworth was born on February 15th 1708 and died on January 10th 1794.

Larynx—Diphtheria

P. 1243. The larynx and adjacent parts of an adult, who died with "putrid sore throat". The mucous membrane lining the larynx is swollen, and a thin imperfect layer of diphtheritic membrane is deposited loosely on its inner surface.

Larynx—Tuberculosis

P. 1244. A larynx, with the base of the tongue and adjacent parts. The mucous membrane of nearly the whole interior of the larynx is superficially and irregularly ulcerated. The vocal cords exhibit nodular infiltration and the ary-epiglottic folds are infiltrated and ulcerated, especially in the position of the left arytenoid cartilage, which appears to have been destroyed. Deep ulcerations extend into the substance of the tonsils.

Larynx—Syphilis

P. 1245. A larynx, part of the trachea, and the base of the tongue, with several probably syphilitic ulcers. Immediately below each vocal cord there is a large ulcer. That on the right side extends below the whole length of the ligament; it is crateriform and deep, with sharp, well-defined borders, and contains a mass of sloughing tissue in its deeper part. That on the left side is more superficial, and has, for the most part, a smooth base, as if partially cicatrised. The adjacent mucous membrane is much swollen; and over the left arytenoid cartilage it is deeply wrinkled from irregular ulceration and contraction. There is a small crescentic ulcer near the left margin of the epiglottis, and a deep ulcer in the right tonsil.

Larynx—Carcinoma

P. 1246. A larynx, with the thyroid gland. Growing from the entire length of the vocal cords and ventricular bands are thickly clustered papillomata.

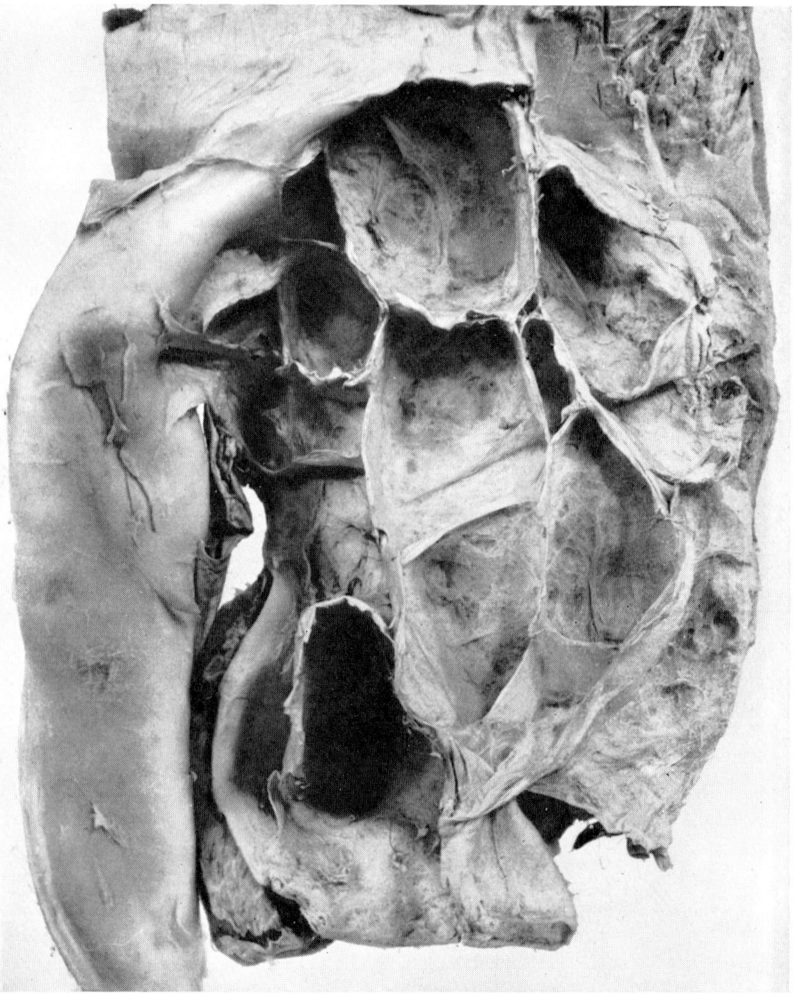

FIG. 17
Part of a human lung, showing recent exudate resulting from pleurisy. Specimen 1249.

Face p. 173

XXI

Lung

P. 1247-1265

Lung—Bronchial Casts

P. 1247. Numerous small masses of a firm brown substance, many of them of rounded form. They are probably casts of bronchiectatic cavities.

Lung—Pleural Adhesions

P. 1248. Portion of the base of a lung with the corresponding part of the diaphragm, injected through the pulmonary artery, the two being united by adhesions of connective tissue as a result of past pleurisy.

Lung—Fibrinous Pleurisy

P. 1249. Part of a lung, on the surface of which there is an abundant, recent fibrinous exudate. The exudate is arranged in two layers, the outer of which has been reflected between which is a space $\frac{1}{2}$ inch in width. This space is divided by partitions into several large cavities, which contained serous fluid.

Lung—Emphysema

P. 1250. Portion of a lung of a Negro, exhibiting a general slight enlargement of the air-cells, with, apparently, some thickening of their walls. Immediately beneath the pleura, also, there are numerous small thinly walled vesicles, of a rounded form, with one or more sides flattened, the larger measuring about $\frac{1}{10}$ inch. The pleura is thickened. Other portions of the lungs contained masses of calcareous matter.

"I have often," says Mr Hunter, "seen such vesicles (containing air) on the edges of the lungs; but these may be supposed to be a kind of aneurismal air-cells filled from the trachea, and are circumscribed and impervious, so that, in the state we find them, they have no communication with the external air." *Works*, vol. iv, p. 98.

P. 1251. Another portion of the same lung, exhibiting similar changes.

P. 1252. Another portion of the same lung.

P. 1253. Another portion of the same lung.

P. 1254. Another portion of the same lung.

Lung—Miliary Tuberculosis

P. 1255. Portion of a lung affected with disseminated miliary tuberculosis. It is studded with innumerable miliary nodules and the intervening lung-tissue is emphysematous.

Tuberculosis

P. 1256. Part of a lung minutely injected. In the section is seen a caseous area, the central part of which has softened and formed a small cavity immediately beneath the pleura. The pleura around is slightly thickened and there is a localised adhesion with the parietal pleura, a portion of which adheres to the specimen.

P. 1257. A piece of lung, apparently from a Bullock, extensively affected with tubercle. Almost the whole of the lung-tissue is replaced by rounded caseous masses embedded in fibrous tissue. Projecting from the pleural surface are numerous sessile and pedunculated tuberculous masses.

Microscopic examination of the lung shows masses of encapsuled caseous material surrounded by compressed lung-tissue. The pleural nodules consist of fibrous tissue containing miliary and conglomerate tubercle.

P. 1258. Portion of the lung of an Ox. From the pleura there hang large numbers of spheroidal nodules. Many occur singly; others, in grape-like clusters.

Microscopic examination of one of the nodules reveals the histological changes characteristic of tuberculosis: the necrotic tissue is in process of calcification. The whole of the pleura was similarly diseased.

P. 1259. Part of the lung of an Ox, the proper substance of which has been completely replaced by a tissue composed of small calcareous granules lying in a fibrous matrix. The pleura, which is thickened, exhibits several rounded nodules. The lesion is probably the result of bovine tuberculosis. There were several similar masses in the lungs.

Microscopic examination shows the presence of large tracts of fibrous tissue in which are islands of calcareous deposit. Between the fibrous tracts there is a loose fibrillated tissue, infiltrated with lymphocytes among which are isolated giant cells, but no true tubercle-systems.

Lung—Calcification

P. 1260. A collection of small particles of calcareous matter, from the lungs of an Ox. They have probably formed in the lesion of bovine tuberculosis.

P. 1261. A nodulated mass probably resulting from calcification of a tuberculous caseous focus, from the lung of a Turkey.

P. 1262. Several small round masses probably resulting from calcification of a tuberculous caseous focus, from the lungs of a Camel.

Lung—Hydatid Cyst

P. 1263. Part of a Lion's lung, containing a hydatid cyst in its substance. A section has been made showing the cyst proper surrounded with a thin adventitious cyst composed of fibrous tissue.

P. 1264. A lung of some small animal, attached to the surface of which there is a globular fibrous cyst, containing a hydatid. Several smaller cysts project from the convex surface of the organ.

P. 1265. A lung of some small animal, from the surface of which there projects a hydatid cyst, measuring about $\frac{3}{4}$ inch in diameter, through the transparent wall of which a smaller cyst can be recognised in its interior.

XXII

Brain, Meninges, Choroid Plexus

P. 1266-1288

Brain—Abscess

P. 1266. A section of the anterior part of a cerebrum, in the left hemisphere of which is the cavity of a large abscess. It has irregular and coarsely granulated walls, and opens anteriorly by an orifice, $\frac{1}{2}$ inch in diameter, through the surface of the brain, and the adjacent and adherent dura mater. Fibrous exudate is deposited on the external surface of the dura mater around the aperture. The soft membranes covering the brain around the aperture are free from inflammation. The affected lobe is markedly swollen, probably from oedema.

Brain of Sheep—Hydatid Cyst

P. 1267. Part of the brain of a Sheep, in the substance of which is a cyst, about $\frac{3}{4}$ inch in diameter containing the shrunken remains of the proper wall of a hydatid cyst.

P. 1268. A sagittal section of the brain of a Sheep, passing through the left cerebral hemisphere. It exhibits a cavity more than 1 inch in diameter between the cerebrum and cerebellum, in which a hydatid was contained. The wall of the cavity is formed by a very thin membrane; the adjacent parts of the brain are hollowed out and pressed aside, but their texture appears unaltered.

P. 1269. The brain of a Sheep, with a cyst in the right hemisphere of the cerebrum measuring about 2 inches in diameter. The wall of the cyst, which contained a hydatid is in part smooth and in part coarsely granulated; at its upper part the cyst is bounded by the pia mater alone.

P. 1270. Part of the skull of a *giddy* Sheep, in which the osseous tissue has completely disappeared in two round areas about $\frac{3}{4}$ inch in diameter, consequent on atrophy due to the pressure of hydatids. The dura mater is adherent to the pericranium, and they together completely close the holes.

Brain—Atherosclerosis

P. 1271. The arteries of a brain affected with diffuse arteriosclerosis. The larger branches especially the vertebrals and basilar, are abnormally rigid from thickening of their walls, and are somewhat dilated.

Dura Mater—Inflammation following Trephine

P. 1272. Portion of dura mater, from a patient who died in St. George's Hospital after being trephined.

The trephine cut an almost complete circle in the dura mater, and both around and within the wound the outer surface of the membrane is coated with fibrinous exudate. On its inner surface, also, fibrinous exudate is deposited in smaller quantity. There are small spicula of bone near the side of the longitudinal sinus, resulting from ossification of the fibres of the dura mater.

P. 1273. Portion of dura mater, on the external surface of which fibrinous exudate is deposited, especially on two circular spots over which the trephine was applied. Its inner surface is blood-stained.

P. 1274. Another portion of the same dura mater.

P. 1275. A portion of dura mater showing two circular areas, covered with fibrinous exudate, in situations where the trephine has been applied. In each area the dura mater had been opened by incisions.

P. 1276. Portion of dura mater, both surfaces of which are covered with inflammatory exudate. Meningitis followed the application of the trephine the situation, of which is indicated by a ring of exudate.

Dura Mater—Ossification

P. 1277. Portion of dura mater, to the outer surface of which, near the longitudinal sinus, a small plate probably consisting of bone is loosely adherent by a thin layer of fibrous tissue.

P. 1278. Part of a dura mater with small portions of adherent pia-arachnoid. Thin plates of bone have formed in the latter membrane.

P. 1279. Portion of pia mater which is thickened and has on its outer surface, embedded in the arachnoid, a thin uneven plate of bone. Small granular masses of bone have also formed in the pia mater. The (single) vertebral and basilar artery is thickened, and calcified.

"*Winter*, 1764-5—I dissected an old man. He was pretty fat, and the muscles red. I found almost all the arteries in his body ossified; the internal carotid and basilary arteries were almost one bone. Those on the pia mater were ossified in a vast number of places; the tomentum cerebrii was ossified in many places. There was but one vertebral artery in this subject. This man had had a bubonocele, which had closed at the abdomen by a very thin, but pretty firm, union."

Dissections of Morbid Bodies, vol. iii, p. 97, No. 93.

P. 1280. Part of a dura mater, dried. To one surface of the falx cerebri there is attached a limpet-shaped mass of bone about $\frac{3}{4}$ inch in its longer diameter.

P. 1281. A portion of dura mater, with the falx cerebri dried. Several portions of bone, having an unusually fibrous or fasciculated and pointed form, are attached to the arachnoid surface of the dura mater, near the margins of the falx.

Dura Mater—Meningioma

P. 1282. Portion of dura mater, on the inner surface of which there is a hemispherical tumour, nearly 1 inch in diameter, and $\frac{1}{2}$ inch in depth. The surface of the growth has a reticulated fibrous appearance.

Microscopic examination shows the growth to be an meningioma.

Dura Mater—Secondary Carcinoma

P. 1283. A portion of dura mater and falx cerebri. Growing from the lower margin of the falx is an oval tumour nearly 2 inches in its chief diameter. Both the surface and section of the tumour have a finely spongy texture and in the section the tumour substance appears in places to radiate from the divided falx which traverses its centre.

Microscopic examination shows the growth to be a carcinoma, some of the cells of which are elongated and more or less columnar in form.

The seat of the primary growth is unknown.

P. 1284. Part of a dura mater, on the outer surface of which there is a thin layer of new growth, $3\frac{1}{2}$ inches in extent. On the inner surface there is in places a similar, but much smaller deposit, and at one spot the pia mater and brain are adherent.

Microscopic examination shows the growth to consist of small alveoli containing masses of polyhedral cells. The growth is probably a secondary carcinoma.

P. 1285. Part of a dura mater, on the outer surface of which there are several large, irregular, nodulated, and flattened masses of new growth. Similar, but smaller masses have also been formed upon its inner surface.

Microscopic examination shows the tumour to be a carcinoma consisting of a stroma of wavy connective tissue bundles, between which are numerous alveolar spaces filled with cells, mostly spheroidal in shape, among which are calcareous spherules in places.

Choroid Plexus—Cyst

P. 1286. Two choroid plexuses, in each of which there are a certain number of small serous cysts.

P. 1287. The choroid plexuses of the lateral ventricles, some of the fringes of which are dilated into small cysts with clear contents.

Choroid Plexus—Meningioma

P. 1288. A choroid plexus, in which there is a small tumour measuring $\frac{1}{2}$ inch in its longer diameter.

Microscopic examination shows the tumour to be a meningioma containing laminated spherules and numerous granules of calcareous matter (psammoma).

XXIII

Spinal Cord, Peripheral Nerves, Eye

P. 1289-1296

Spinal Cord—Compression

P. 1289. "A portion of the medulla spinalis, which had been compressed."

MS. Catalogue.

The dura mater and, at the upper part of the specimen, the arachnoid, has been laid open. The substance of the cord appears to be softened throughout and there are localised constrictions in two situations, at which the cord must have been completely severed.

P. 1290. A portion of spinal cord, compressed, near the cauda equina. The extremity of the cord is softened and disorganised.

Spinal Cord—Softening

P. 1291. "A portion of medulla spinalis, in which the natural structure is indistinct."

MS. Catalogue.

The central part of the cord has undergone softening. A longitudinal incision has been made along one surface and the cord opened out. There is also considerable softening of the peripheral part of the cord, immediately beneath the arachnoid, which is thickened.

Peripheral Nerve—Amputation Neuroma

P. 1292. Portion of a nerve from the stump of a thigh of a man after amputation, exhibiting a bulb-like enlargement of its cut extremity.

P. 1293. Portion of one of the nerves from the same stump. The end is enlarged into a flattened spherical swelling 1·5 cm. ($\frac{3}{5}$ inch) in diameter, which has resulted from the growth of nerve fibres from the ends of those divided in the amputation.

P. 1294. A section of a nerve, from the same amputation stump. Most of the bundles of nervous fibres appear to cease abruptly at the commencement of the bulb-like enlargement; one can be traced for a short distance into its substance. The section of the enlargement itself exhibits a uniform, firm and compact texture, like that of a cicatrix of the integuments.

P. 1295. Another nerve, from the same stump, in which several bundles of nervous fibres have been traced by dissection, diverging at the lower end of the nerve, and passing for a short distance into the bulb-like enlargement.

Eye—Ossification following Inflammation

P. 1296. The eye of a blind man, in which large plates of bone have been formed in the choroid, probably after suppurative inflammation.

The lens is completely calcified.

XXIV

Kidney, Ureter

P. 1297-1341

Kidney—Congenital Hypoplasia

P. 1297. "The kidney become very small, and the ureter much enlarged. The patient never had any complaint in this kidney; the other was of the natural size."

MS. Catalogue.

The kidney is scarcely more than two inches long; it is lobed, and looks shrivelled; the size of its blood-vessels is proportionally diminished; the pelvis and ureter are dilated, but not structurally diseased.

P. 1298. A shrunken lobulated kidney which has completely lost its characteristic outline. The ureter is small, the pelvis is dilated, and the distended calyces project from the surface as thick-walled sacs. The condition probably represents a terminal infection in a hypoplastic kidney.

Kidney—Congenital Cystic Disease

P. 1299. Section of a kidney, enlarged to four or five times its natural size by the formation of numerous cysts within it and upon its surface. The cysts vary in size up to an inch and a half in diameter: the largest are in the central parts of the kidney; their original form appears to have been spherical, but where they are closely crowded they have become irregular and angular. Their walls are formed of distinct thin membrane, polished on its internal surface, and externally intimately connected with the remains of the substance of the kidney. Their contents appear various; some are filled with transparent, probably serous fluid; some with a dark yellowish or brown and viscid substance; some with a turbid, nearly solid, material. The original substance of the gland is in some parts wholly removed; in other parts small portions remain intervening between the cysts, and appearing healthy in their texture.

P. 1300. The other section of the kidney last described. The renal capsule is preserved with it, and appears healthy.

Kidney—Cortical Cyst

P. 1301. A kidney with a cyst embedded in its outer surface. The cyst is oval, about 3 inch in its greatest diameter, and situated upon the anterior surface of the kidney. The kidney is of natural size, but slightly granular. Its blood-vessels have been partially injected with a white fluid.

P. 1302. A kidney, with a cyst nearly 4 inches in diameter attached to its surface, but scarcely at all embedded in it. Its interior is smooth.

P. 1303. A thin-walled membranous cyst nearly 2 inches in diameter, half embedded in the substance of a kidney, and half projecting from its surface. That part of the wall of the cyst which is embedded in the kidney is fasciculated and folded in consequence of the unequal resistance offered to its growth. The adjacent surface of the kidney is granulated.

Kidney—Abscess

P. 1304. Part of a kidney, in the tubular substance of which is the cavity of a small, irregularly walled abscess.

From a man who died at St. George's Hospital after the operation for the stone.

P. 1305. A portion of the other kidney of the same patient. The surface of one of the papillae is incrusted with calcified inflammatory exudate.

Kidney—Multiple Abscesses

P. 1306. Section of a kidney, enlarged and exhibiting numerous small oval abscesses scattered irregularly through its substance and immediately beneath its surface. The pelvis, calyces, and ureter are dilated but their texture appears healthy.

Kidney—Perinephritic Abscess

P. 1307. A kidney, surrounded with a sac with walls ¼ inch thick, hard, dense, and apparently composed of the capsule and surrounding tissue consolidated by inflammation. The interior of the sac is lined with a layer of granulation tissue and in one situation, opposite the hilus, it is adherent to the surface of the kidney. The kidney itself is contracted, granular, and in part, covered with lymph.

Kidney—Tuberculosis

P. 1308. Section of a kidney, in which the extremities of some of the pyramids have been destroyed by irregular ulceration, in consequence of tuberculous disease. There are a few small tuberculous deposits in other parts of the kidney; and beneath the mucous membrane of the pelvis there are several isolated tubercles, some of which have softened and ulcerated by small apertures through the surface.

P. 1309. Section of a kidney, apparently from a child, in which, in the place of nearly all the natural texture, there are three large cavities roughly lined with tuberculous granulation tissue. The tissues around the kidney are thickened and indurated, and a calculus is imbedded in it near its surface.

P. 1310. Section of a kidney, similarly diseased, probably from the same child, for the size and form of both the kidney itself and the cavities in it are exactly alike.

P. 1311. Sections of a kidney and of a renal capsule, "from a butcher whose smelling was remarkably acute." The substance of the kidney is atrophied, shrunken, and contracted, as if by cicatrisation, especially in its middle part, where it is not altogether more than ½ inch in thickness. At this part, also, there is a deposit of white substance, probably thickened and dried tuberculous matter, in the kidney; and in other parts there are large serous cysts. An unusual quantity of fat is accumulated around the pelvis and its divisions. The renal capsule is enlarged and firm.

P. 1312. A kidney, of which the lining membrane of the pelvis and ureter is thickened, granulated, and in parts superficially ulcerated. The substance of the kidneys appears healthy.

P. 1313. A kidney enlarged, and with the lining membrane of the pelvis, calyces, and ureter dilated, thickened, very coarsely granulated, and in many places covered with tuberculous matter. On many parts, also, of the lining membrane there are small, superficial, oval and circular ulcers.

Kidney—Hydatid Cyst

P. 1314. Sections of a kidney, probably from a Sheep, the interior of which is full of acephalocyst hydatids, some with thin and pellucid, others with thicker and opaque, membranes. In consequence of their growth, the glandular substance of the kidney is reduced to a narrow layer. Neither the form nor the size of the kidney appears altered.

P. 1315. A section of a Sheep's kidney full of large acephalocyst hydatids.

P. 1316. The other section of the same kidney.

P. 1317. The kidney of a Sheep, in which is embedded a hard globular mass, an inch and a half in diameter, and composed of concentric layers of alternately yellowish and white substances. It probably consists of the debris of acephalocyst hydatids, which died and burst, and of which the membranes were then rolled up and compressed and became the seat of calcareous deposits. The structure of the kidney appears healthy.

Kidney—Pyelonephritis

P. 1318. Section of a kidney, with part of the ureter. The coats of the ureter are thickened, and its mucous membrane is granulated. About 1 inch from the pelvis the ureter makes a very abrupt turn, and one of its walls is folded, almost like a valve stretching across its canal and hindering the passage of fluid through it.

Kidney—Hydronephrosis

P. 1319. Two kidneys, with portions of their ureters. One of the kidneys is reduced by atrophy to half its natural size; its glandular substance is nearly all removed; its pelvis, which contained a large calculus, is dilated; and so is its ureter, in which a smaller calculus was lodged. The other kidney is, in external dimensions, much enlarged; and its glandular substance is little, if at all, diminished, though expanded around the pelvis and calyces, which contained numerous calculi, and are dilated even more than those of the kidney first described.

P. 1320. Section of a Sheep's (?) kidney, with its ureter. The ureter, pelvis, and all its branches are very much dilated; and the glandular substance of the kidney is so expanded and atrophied, that it forms a single thin-walled sac, with shallow recesses formed by the projection of the remains of the calyces.

P. 1321. Section of a kidney, of which the ureter, pelvis, and infundibula are much enlarged, and nearly all the glandular substance is removed. The organ is thus reduced to a thin-walled multilocular sac, of rather less size than the healthy kidney. Its capsule is thickened and indurated.

P. 1322. Section of a kidney, of which the pelvis, calyces, and proper substance are dilated, but not changed in texture. In two places, the flattening and pressing back of the papillae are shown, as in the preceding specimens; each papilla, instead of projecting into the extremity of the calyx, forms but a slight prominence beneath its distended mucous membrane.

P. 1323. A hydronephrotic kidney in which the dilatation of the pelvis and other parts, and the removal of the substance of the kidney are very advanced.

Works, pl. xiv.

P. 1324. A hydronephrotic kidney. In consequence of stricture of the urethra, the ureter, pelvis, and calyces are all enlarged; the papillae of the kidney are no longer prominent, and much of its proper substance has been lost by atrophy. The external form and size of the kidney are not altered.

P. 1325. A kidney, with the surrounding substance. The glandular substance of the kidney is reduced to a layer which is not anywhere more than one-third of an inch thick; and the papillae are very small. The surface of the kidney is seamed and puckered, but not granulated. The pelvis and calyces are all much dilated, but they are healthy in texture. The capsule of the kidney and a layer of fat nearly half an inch thick, surrounding it and the pelvis, appear to have coalesced so as to form one distinct sac.

From a man who died at St. George's Hospital after operation for the stone.

Ureter—Hydroureter

P. 1326. Section of a kidney, showing, as is usual in atrophy, a greater than natural accumulation of fat around the pelvis, infundibula, and calyces. The kidney itself is rather reduced in size, its surface is slightly lobed, and there are smooth-walled serous cysts in its interior. The pelvis and calyces are dilated; and the ureter is dilated and tortuous and its coats are thickened and indurated.

P. 1327. Part of an ureter, which in consequence of stricture of the urethra, is dilated to nearly 1 inch in diameter. Its walls are as thick as those of a healthy ureter, and are healthy in their texture.

P. 1328. Two kidneys, with their pelves and ureters, and the urinary bladder, injected with wax and dried. The pelves and ureters are greatly dilated, and the kidneys appear sacculated, but are not enlarged.

Kidney—Calculi

P. 1329. A kidney, with the pelvis and its branches dilated, and some small calculi adhering to the papillae as if embedded in their substance.

P. 1330. The kidney of a Dog, with numerous small calculi in the pelvis and calyces.

P. 1331. A kidney of a man, the pelvis of which was nearly filled with the numerous small white calculi and particles which now lie at the bottom of the bottle. The lining membrane of the pelvis is thickened and flocculent; the surface of the kidney is irregularly granulated.

P. 1332. "The left kidney of a child, who died at St. George's Hospital, where a stricture was suspected. The pelvis and infundibula are enlarged and lined by a layer of coagulable lymph. Calculi are lodged in the infundibula."

MS. Catalogue.

P. 1333. "The right kidney of the same child, where the infundibula are also enlarged and ulcerated, so as to communicate with the cavity of the abdomen."

MS. Catalogue.

In both kidneys, together with the enlargement of the pelvis and infundibula, the glandular substance is much reduced in many places. At the upper end of the right kidney it appears to have been ruptured.

P. 1334. Portion of a kidney, with three large calculi filling dilated and ulcerated calyces.

P. 1335. The kidney of a lad who died at the London Hospital. The pelvis, its branches, and the commencement of the ureter are much dilated, and they are all nearly filled by two large calculi, which have grown in the branched form of the parts containing them. The substance of the kidney is in several places almost wholly removed, and in all the parts much atrophied.

P. 1336. A kidney, of which the ureter, near the pelvis, is completely obstructed by a large calculus exactly adapted to its form. The pelvis and its branches are thickened and dilated, and a calculus similar to that in the ureter is lodged in one of the calyces. The substance of the kidney is atrophied.

P. 1337. A kidney, of which the pelvis and the beginning of the ureter are filled by a calculus. All the branches of the pelvis are exceedingly dilated; so that the kidney has acquired the form of a large multilocular sac, surrounded by a thin layer of glandular substance. The lining membrane of the sac is healthy in its texture, and thicker than that of the calyces in their natural state.

P. 1338. A similar specimen; but the kidney is distended to a much larger size, measuring 8 inches in length and $4\frac{1}{2}$ inches in breadth. The lining membrane of the enlarged calyces, pelvis, and ureter is thickened and inflamed. The capsule of the kidney and the adjacent tissues are thickened, hardened, and consolidated.

P. 1339. The remainder of the ureter, and a portion of the bladder from the same patient as the kidney last described. The ureter is dilated; its coats are thickened and its interior is irregularly ulcerated. The bladder is inverted, and presents several circular sharp-bordered ulcers in an otherwise healthy mucous membrane.

*Kidney—Wilms' Tumour**

P. 1340. The left kidney of a child, of the natural size, and apparently healthy externally and in section; but occupying its hilum is a rounded myo-sarcoma, 1 inch in

* Wilms, Max (1867-1918), German surgeon. *Die Mischgeschwülste der Niere.* Leipzig, 1899.

diameter, which completely obliterates its pelvis. The ureter passes out between the kidney and the posterior surface of the tumour. The section of the tumour presents the appearance of interlacing bands, and is of a yellowing or faint brownish-white colour. The new growth is sharply defined from the kidney-substance by a layer of connective tissue, but the capsule of the kidney is continued over it.

Histologically the tumour consists of richly cellular connective tissue with round-celled sarcomatous areas in which are distributed large numbers of striated muscle fibres. The cysts present in the tumour may be traced as arising from local degeneration as they have no epithelial lining.

P. 1341. A tumour, 7 inches in length and 5 inches in width, connected with the right kidney of the same patient. Its section shows interlacing bands of connective tissue, intermixed in places with a softer material, and thickly dotted in its lower half with cysts, none of which exceed one third of an inch in diameter. The hilum and inner surface of a small, ill-developed kidney are loosely connected by areolar tissue to the surface of the tumour on the observer's right side.

In minute structure the tumour resembled the preceding, except in that the muscular fibres were more abundant, and formed fasciculi either parallel to each other or interlacing.

XXV

Bladder

P. 1342-1355

Bladder—Hypertrophy

P. 1342. Portion of the fundus of a bladder, the muscular coat of which is $\frac{1}{2}$ inch thick, though not contracted. The external fasciculi of muscular fibres are very strongly developed, and are prominent in curved and interlacing ridges. The mucous membrane is closely and deeply corrugated, but healthy in texture.

P. 1343. Section of a bladder, with the prostate gland. In consequence of stricture of the urethra, the muscular coat of the bladder is greatly increased in thickness; and its fasciculi are very prominent beneath the mucous membrane, which itself appears very thin and soft.

Bladder—Cystitis

P. 1344. Portion of a bladder, of which, in consequence of stricture of the urethra, the muscular coat is thickened. The mucous membrane has been inflamed, and is nearly covered with a layer of inflammatory exudate with particles of calcareous matter imbedded in it. Some small calculi are also enclosed in pits of the mucous membrane depressed into the meshes of the muscular fasciculi.

P. 1345. A bladder, in which there had been a calculus. Its mucous and muscular coats are thickened and indurated; the former is superficially ulcerated; its internal surface has a dark granulated aspect, and thin layers of exudate mixed with calculous particles adhere to the surfaces of its ridges.

Bladder—Ulceration

P. 1346. A bladder, containing a large calculus. The bladder is enlarged and remarkably elongated; its mucous membrane is ulcerated and shreddy; the hypertrophied muscular fasciculi stand out in prominent ridges upon its surface, forming a kind of coarse network, through some of the meshes of which sacculi of the other coats had begun to form. The prostate gland is enlarged to nearly three times its natural size. The calculus appears to have been held above it in the elongated and contracted bladder.

Bladder—Calculi

P. 1347. A bladder and prostate gland, together with the ureters and kidneys. The upper third of the bladder is deeply fasciculated, and appears contracted; it is separated by a broad and prominent muscular fasciculus from the lower two-thirds, the walls of which are dilated and not at all fasciculated. The lining membrane of the lower two-thirds is generally ulcerated, and is covered with a layer of exudate in which numerous small white angular calculi are embedded and in some places, form a continuous incrustation. In the upper third there are several small calculi in the pits between the projecting fasciculi of the muscular coat, but the mucous membrane appears healthy. The prostate gland is enlarged, and the lining membrane of the urethra within it is ulcerated and incrusted with calculous matter. The ureters, pelves, and calyces are enlarged. Much of the substance of the kidneys has disappeared, but what remains is healthy.

Bladder—Diverticulum

P. 1348. A bladder, in which a wide and deep pouch is formed by a partial dilatation of all the coats directly behind the prostate gland. The pouch is wider at its communication with the bladder than at any other part; it is directed rather towards the left side; it is such as one as may have been caused by the lodgment of a calculus. The walls of the bladder are thickened and indurated.

P. 1349. A bladder and prostate gland. In consequence of stricture of the urethra, the muscular coat of the bladder is hypertrophied; and in the posterior wall there are, as in the preceding specimen, many pouches of the mucous membrane, unusually regular in form and size, pushed out between the muscular fasciculi, but not protruded far enough to be prominent externally. The prostate is enlarged.

P. 1350. A bladder, distended and dried. There are two sacculi, probably formed of the mucous membrane protruded through the muscular coat, in the posterior wall; and one similarly formed in the middle of the anterior wall. This last is hemispherical, and measures one and two-thirds inches in diameter; its orifice is circular, and 1 inch in diameter.

P. 1351. A bladder, at the posterior part of which there are two sacculi, formed by partial dilatation of all the coats, and each containing a rough irregularly shaped calculus. The orifices by which the sacculi communicate with the cavity of the bladder are contracted. The muscular coat of the bladder is $\frac{1}{2}$ inch thick; its cavity is contracted to $1\frac{1}{2}$ inch in diameter; its mucous membrane indurated and thickened; and the prostate gland is slightly enlarged.

Bladder—Carcinoma

P. 1352. A bladder, of which all the coats are perforated by a small ragged ulcer, just above and outside the right ureter. In other parts it appears healthy.

P. 1353. "A diseased bladder, from which was a bleeding. It has some of the coagulated blood adhering to it, and some of the loose coagula lying at the bottom of the bottle." (*MS. Catalogue.*) The mucuous membrane is nearly all removed, as if by sloughing, from the lower and posterior part of the bladder; its remains are ragged and flocculent; the exposed subjacent tissues are neither thickened nor indurated. The prostate is somewhat enlarged.

P. 1354. Fragment of the papillomatous tumour from the previous specimen.

Bladder—After Lithotomy

P. 1355. A bladder and penis, laid open from the left side, and part of a perineum. The urethra is of its natural size, but two fistulous passages (indicated by bristles) lead from it to the perineum. One of these commences directly in front of the prostate gland, beneath which it passes for one inch backwards; the other, beginning at the bulb of the urethra, communicates at once with a large smooth-walled cavity, which has a long oval opening in the perineum. There is a large and partially ulcerated protrusion of the wall of the bladder close by the right ureter, as if a calculus had lain there. It is probable that lithotomy was performed some time before death, and that the wound of the urethra only partially healed.

XXVI

Vesiculae Seminales, Prostate

P. 1356-1367

Vesiculae Seminales of Lemur—Dilatation

P. 1356. The bladder, the vesiculae seminales, prostate, and Cowper's glands of a Lemur. One of the vesiculae is enlarged.

Prostate—Hypertrophy

P. 1357. Part of a bladder, with an enlarged prostate gland. The degree of enlargement is slight, but nearly uniform, affecting equally both the lateral and middle lobes. The middle lobe is exposed by dissection from behind; and it forms a slight swelling in the neck of the bladder, oval, and elongated from before backwards. The section of the anterior and middle portion of the gland shows that this also is much increased in thickness. Bristles are placed in the ureters and ejaculatory ducts.

P. 1358. A bladder, prostate gland, and adjacent parts. The prostate gland is uniformly enlarged to more than twice its natural size, and the part of the urethra within it is very deep and narrow. The middle portion of the gland is elevated like a transverse barrier with a smoothly rounded, level, upper margin across the neck of the bladder. The mucous membrane of the bladder is in several places pushed-out in sacculi of various size and depth between the bundles of the hypertrophied muscular coat, which are peculiarly strong above and behind the enlarged prostate. Drawing No. 119.

P. 1359. The kidneys from the same case showing marked dilatation of the pelvis and calyces with corresponding atrophy of the renal tissue. One of the kidneys contains a rounded cortical tumour nearly 1 inch in diameter.
Drawing No. 121.

P. 1360. Part of a bladder, with a uniformly enlarged prostate gland. The middle portion of the gland appears wider than is natural, but forms no prominence

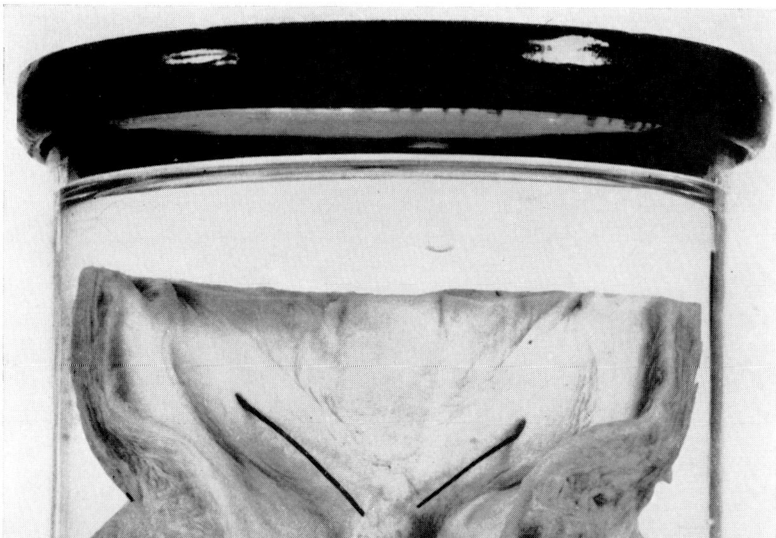

FIG. 18
Part of the bladder with enlarged prostate gland. Specimen 1357.

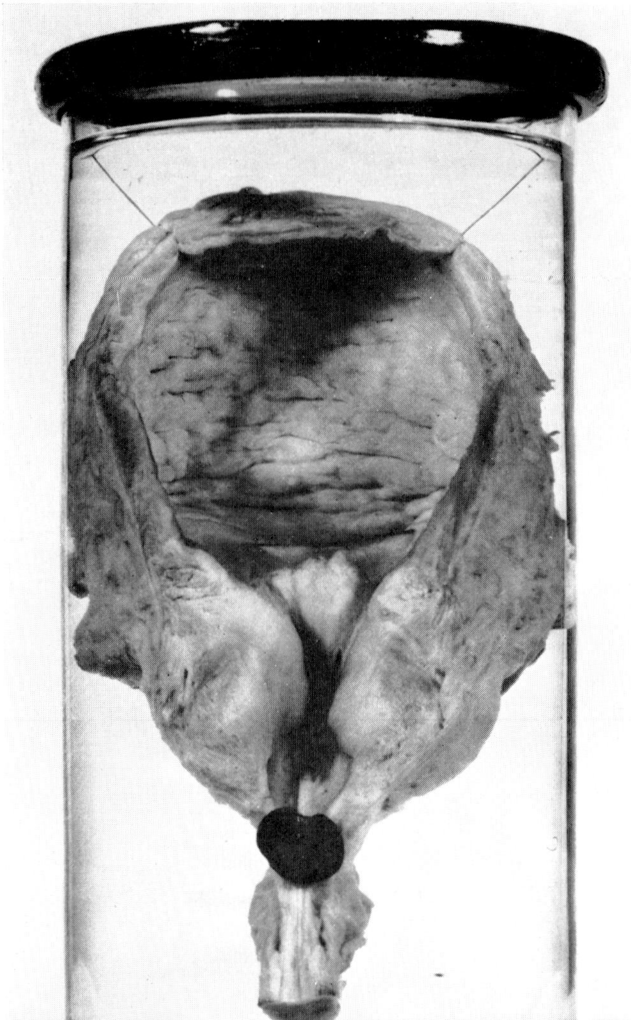

Fig. 19
The bladder and adjacent parts from a patient who died as a result of retention of urine caused by a calculus in the urethra. Specimen 1367.

in the neck of the bladder. Numerous cavities in the prostate open by wide smooth orifices into the urethra. One of them is in the situation of the right ejaculatory duct. The others were probably made with catheters. The prostatic part of the urethra appears to have been large, and little, if at all, laterally compressed. There is a circumscribed cavity behind the lower part of the bladder, with rough granulated walls—the remains, probably of an abscess between the bladder and rectum. The muscular coat of the bladder is hypertrophied, and the transverse band of fibres between the ureters is peculiarly large and prominent.

P. 1361. Part of a bladder, with an enlarged prostate. Numerous deep sacculi of the mucous membrane of the bladder are pushed-out between the bundles of the hypertrophied muscular coat. The prostate is enlarged to three or four times its natural bulk, and its section presents a coarsely fibrous aspect, with bright or shining white fibres curving and circling in a pale firm homogeneous basis, much like a fibrous tumour. The right lobe is by one half larger than the left, and the middle lobe is prominent in the bladder. The prostatic part of the urethra is deep and narrow, and curved in adaptation to the enlargement of the right lobe of the prostate. The caput gallinaginis is prominent and sharp-edged. The ureters are dilated.

P. 1362. Part of a bladder, with an enlarged prostate gland. The enlargement has taken place to a much greater extent in the left than in the right lateral lobe of the prostate. The middle lobe is scarcely prominent; the left lobe forms a low, smooth, convex elevation in the neck of the bladder. The vesical orifice of the urethra, in adaptation to the change of form in the prostate, has the shape of a narrow crescent, with its horns $\frac{3}{4}$ of an inch apart and turned to the left side. This orifice is also much nearer to the anterior than to the posterior part of the bladder, in consequence of the posterior middle portion being much more increased than the anterior middle portion is. The muscular coat of the bladder is hypertrophied; in other respects it is healthy.

P. 1363. Part of the bladder with the prostate and adjacent parts. The prostate is slightly enlarged, especially the middle lobe which projects into the bladder. The bundles of muscular fibres passing from the orifices of the ureters to the posterior part of the prostate are remarkably large.

P. 1364. Part of a bladder, with an enlarged prostate gland. The enlargement has taken place chiefly in the middle lobe, which is raised in a broad, thick transverse ridge or bar behind and below the vesical orifice of the urethra. The anterior surface of this ridge forms nearly a right angle with the rest of the prostatic part of the urethra; and its prominence appears the greater from the urethra directly in front of it being dilated into two deep fossae by the sides of the caput gallinaginis. There are small calculi at the orifices of two of the prostatic ducts. The cut surfaces of its lateral lobes display an irregularly fibrous texture, white fibres running with various curves through a pale homogeneous basis; and the cut orifices of numerous slightly dilated ducts contain minute calculi.

P. 1365. Part of a bladder, with the prostate gland enlarged into a spheroidal mass nearly 3 inches in diameter. The middle lobe forms a tumour projecting into the neck of the bladder, upwards of an inch in width, and three-quarters of an inch in depth, and marked on its anterior surface with a deep groove, in which a bougie long rested. The prostatic portion of the urethra has a nearly vertical direction, and is much elongated.

P. 1366. A bladder, prostate gland, and part of the penis. The bladder is very large, its muscular coat is hypertrophied, its mucous membrane thickened and flocculent, especially about the neck, where it is also in many places superficially ulcerated. The prostate gland is increased to nearly 3 inches in width. In the neck of the bladder it projects in four distinct knobbed pyriform lobes, each about 2 inches long and 1 inch in thickness at its larger end, and all tending at their narrow ends towards the caput gallinaginis. The two middle of these lobes are roughly ulcerated on their surfaces. The prostatic part of the urethra is very deep and large, and exhibits several depressions or cavities in its wall, produced probably by the forcible introduction of instruments. The rest of the urethra, so far as it is preserved, is also very large, and has been similarly injured in several places. About the middle of its spongy portion there is a false passage, nearly 2 inches in length, by the side of which a great part of the wall of the urethra appears to have been destroyed.

P. 1367. A bladder, with the prostate gland, part of the penis, &c. The muscular coat of the bladder is thickened; its mucous membrane is healthy. The prostate gland is enlarged to more than three times its ordinary dimensions, and proportionally in all its lobes. The urethra, obstructed by the projection of the middle lobe, and narrowed by the enlargement of the lateral lobes, is completely blocked by a calculus, which is lodged in its prostatic portion just over the caput gallinaginis, and appears to have been tightly grasped by the elastic substance of the prostate.

The patient died in consequence of retention of urine.

XXVII

Urethra

P. 1368-1380

Urethra—Impression of Stricture on a Bougie

P. 1368. Portions of three bougies, near the points of which there are circular impressions, produced by their being passed into annular strictures of the urethra.

P. 1369. Portions of three bougies, with impressions at a distance from their points, produced by stricture.

Urethra—Stricture

P. 1370. Part of a penis, with a stricture at the bulbous and adjacent parts of the urethra. The canal is "irregularly contracted for above an inch in length, owing to its coats or internal membrane being irregularly thickened and forming a winding canal." *Works*, vol. ii, p. 230.

P. 1371. A bladder and penis, laid open from the front and right side. Nearly the whole length of the urethra is diseased; its mucous membrane is thickened, and in many places contracted and puckered, with shining, wavy, fibrous bands variously arranged in or beneath it, looking like the surface of a cicatrix after superficial ulceration. Bristles are introduced into four places, at which the canal of the urethra is more especially narrowed. At the junction of the membranous and prostatic portions, its walls are deeply ulcerated. The texture of the corpus spongiosum is indurated, and like fibrous tissue. The mucous membrane of the bladder is corrugated and granular; its muscular coat is hypertrophied.

P. 1372. A bladder, and part of a penis, laid open from the front. Part of the membranous and the bulbous portions of the urethra are contracted irregularly. An inch and a half of the urethra anterior to the stricture has been preserved, and is dilated to nearly twice its ordinary size; its lining membrane appears to have been inflamed, or superficially ulcerated. The walls of the bladder are thickened and contracted. The dilatation of the urethra anterior to the stricture was, doubtless, produced by bougies long pressed against the stricture and the parts in front of it.

P. 1373. Parts of a prostate gland and of a penis, laid open so as to give a lateral view of the urethra. A stricture appears to have obliterated the posterior part of the bulbous and the greater part of the membranous portions of the urethra; what remains of it is indicated by pieces of bristle. A bougie is placed in a false passage formed by caustic. It commences half an inch anterior to the stricture, and after a course of an inch and a half, re-enters the urethra in its prostatic portion, which is dilated much beyond its natural size.

Urethra—Fistula

P. 1374. A bladder, penis, scrotum, and other adjacent parts. The urethra, and a fistulous passage leading from it to the perinaeum, are laid open from the left side. There is a very narrow annular stricture, nearly closing the bulbous portion of the urethra; a part of its lining membrane is ulcerated, and a white bristle is passed through the most contracted portion. Behind the stricture the membranous and prostatic portions of the urethra are dilated, and the lining of the membranous portion is thickened and puckered. An inch behind the stricture there is a small aperture in the lower wall of the urethra through which a black bristle is passed; it is the orifice of the small fistulous canal already mentioned, which extends with a tortuous course, and gradually enlarging, beneath the integuments; and, after branching, opens in the middle of the perinaeum, and in the posterior part of the scrotum. The lining of the fistula is rough, as if covered with granulations. The muscular coat of the bladder is exceedingly hypertrophied, and its fasciculi project in strong columns beneath the thickened mucous membrane.

P. 1375. A bladder, penis, scrotum, and perinaeum. The bladder and part of the urethra are laid open from the left side, and a bougie is placed in their whole course. The membranous portion of the urethra, and the anterior half of the prostate gland, are involved in a deep and irregular ulceration, which has formed in their place a large oval cavity, like that of an abscess. Portions of whalebone are passed into several apertures in the scrotum and perinaeum, which lead into fistulous canals communicating indirectly with the cavity above described. The bladder is very small, and its coats are thickened. One of the testicles is exposed; it is enlarged, and the cavity of the tunica vaginalis is obliterated by adhesion of its opposite surfaces.

P. 1376. Parts of a bladder and penis, the greater portion of one side of the penis being removed for the exposure of the urethra. A short fistulous passage leads from the urethra through the integument, by the side of the fraenum of the prepuce; it is about a line in diameter, and its edges are smoothly rounded, as if it had long existed. The urethra anterior to the aperture, and for nearly 2 inches behind it, is irregularly contracted, and its surface is fasciculated, as if it had cicatrised. Behind the contraction the urethra is dilated as far as the bulbo-membranous portion, where there is a second slight contraction, and a short false passage in which a bougie is placed. The bladder is contracted; its mucous membrane is thickened, dark, and granulated on its inner surface; and on the right side, near the neck of the bladder, a kind of polypous growth, consisting of groups of small, oval, flattened bodies, is attached to the mucous membrane of the bladder.

Urethra—Periurethral Abscess

P. 1377. A bladder and urethra, laid open from the front. The bladder is contracted, its muscular coat is hypertrophied, its mucous membrane is everywhere superficially ulcerated and covered with patches of lymph and calculous matter. The mucous membrane of the prostatic portion, and of the anterior part of the spongy portion, of the urethra is similarly diseased; the membranous portion and the spongy portion, for two inches anterior to the bulb, have their walls thickened, indurated, covered with lymph, and so irregular that the canal can hardly be traced. By the side of this part bougies are placed in two fistulae or false passages, leading very obliquely towards the perinaeum.

Urethra—Perineal Abscess

P. 1378. Part of a bladder, with the penis, scrotum, and adjacent integuments. Part of the urethra has been exposed from above, and a bougie is passed through its whole length. All the membranous, and a great part of the spongy, portions of the urethra are involved in deep and irregular ulceration, which just in front of the membranous portion, has formed a large aperture, communicating with a cavity in the tissues of the perinaeum. There is a small opening behind the scrotum near the raphé, which leads straight and at once into the cavity. The bladder is very small, and its coats are thickened; but its lining membrane is healthy. There is no trace of the stricture of which it is probable these changes were the consequence; doubtless, as in some of the foregoing cases, it was removed by the ulceration.

Urethra—Malignant Tumour

P. 1379. Section of a penis, in which the areolae of the bulb and of a great part of the corpus cavernosum urethrae, appear to be filled with a growth of epithelial cancer. At the anterior part of the penis there is a large unevenly ulcerated aperture in the urethra, and the whole length of the canal is rough and narrow. Some of the vessels on the back of the penis appear filled with coagula.

Urethra of Ox—Tumour

P. 1380. Part of the bladder and urethra of an Ox. A pyriform growth is attached to the verumontanum, and reaches backwards to the neck of the bladder. It is about 2 inches in length, has a round pedicle nearly 1 inch long and is $\frac{3}{4}$ inch in diameter at its larger end. Its surface is smooth and covered with mucous membrane, like that lining the urethra. The part of the urethra around the growth is dilated, but, as well as the adjacent parts, is healthy in its texture.

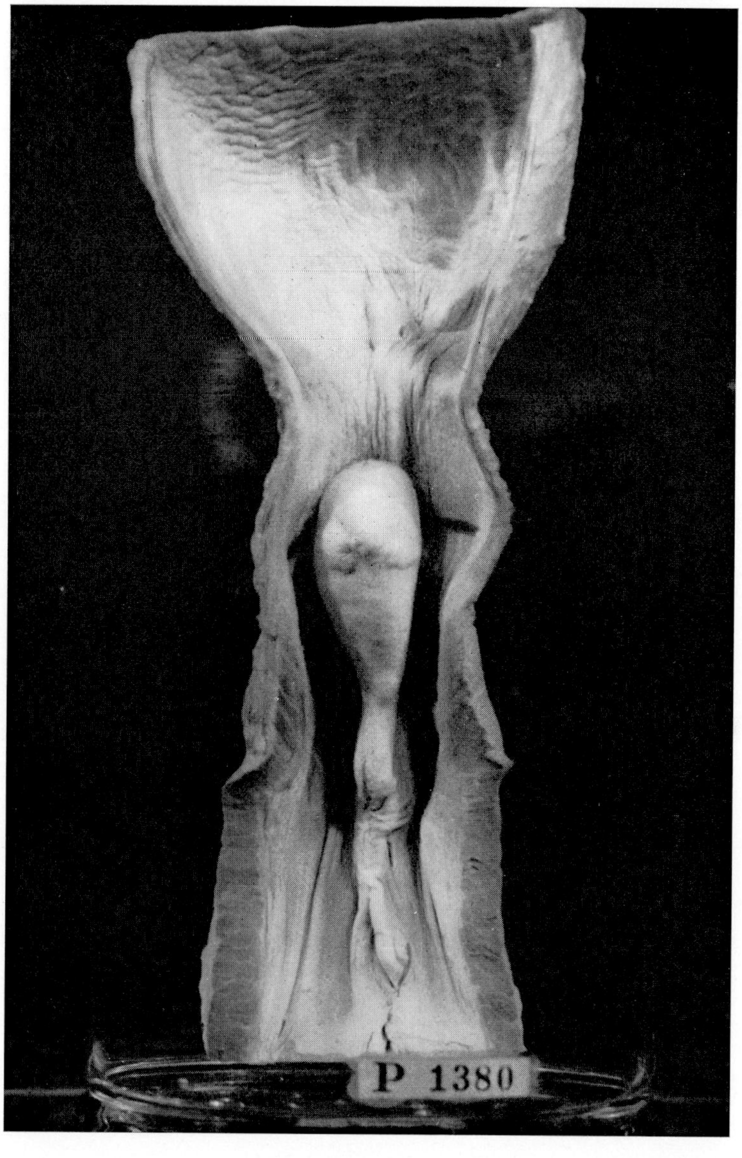

FIG. 20
Part of the bladder of an Ox, together with the urethra in which there is a tumour. Specimen 1380.

XXVIII

Penis

P. 1381-1394

Penis—Oedema of Prepuce

P. 1381. Part of a penis, with phimosis, produced chiefly by inflammation and oedematous enlargement of the prepuce.

Penis—Adherent Prepuce

P. 1382. Part of a penis, in which the prepuce has become completely adherent to nearly the whole of the glans, leaving but a small orifice for the passage of the urine. Hairs have grown on the lower surface of the prepuce, even to its extremity.

See P. 5 on growth of hair.

Penis—Phimosis

P. 1383. Part of a penis, in which the operation of slitting the prepuce, for the cure of phimosis, was performed long before death. The edges of the wound have smoothly healed, but they still appear swollen: the glans is half exposed.

Penis—Preputial Concretion

P. 1384. Part of a penis, with small warts in the interior of the retracted prepuce, and a large collection of the secretion of the glandulae odoriferae adhering around the glans, dried, and hardened into a thick laminated substance like cuticle.

Penis—Inflammatorv Thickening of Prepuce

P. 1385. An enlarged prepuce removed by operation. It forms a solid, irregular, oval mass, about 3 inches in diameter, dense, pale, and compact, and apparently composed of fibro-cellular tissue like that of loose skin, unmixed with fat.

Penis—Ulceration of Prepuce

P. 1386. Part of a penis, in which, after the glans had protruded through the ulcerated prepuce, all the diseased parts healed. The prepuce, reduced to its natural dimensions, and retaining its natural saccular form, has its base attached to the inferior surface of the glans. The skin around the base of the glans is drawn tight; but there does not appear to have been any constriction of the urethra.

Penis—Syphilis

P. 1387. The end of a penis, exhibiting on the inferior surface of the glans two circumscribed depressions, left after the healing of syphilitic ulcers.

P. 1388. Part of a penis, in which a syphilitic ulcer, extending through the prepuce, has healed.

P. 1389. Part of a penis, from which the prepuce has been completely removed by syphilitic ulceration. The ulceration has smoothly cicatrised, except on a part of the corona glandis, to which it appears to have spread from the prepuce.

P. 1390. The end of a penis, in which a great portion of the glans has been destroyed by syphilitic ulceration. The ulceration appears to have been nearly healed when the patient died.

P. 1391. Part of a penis, in which the fraenum of the prepuce and the adjacent part of the wall of the urethra were destroyed by syphilitic ulceration. The ulcers have healed, but the aperture in the urethra remains: a piece of whalebone is placed in it.

P. 1392. Part of a penis, in which an ulcer, by the side of the fraenum of the prepuce, has penetrated the urethra close to its external orifice. A bristle is tied round the remaining partition between the orifice of the urethra and that of the ulcer.

Penis—Papilloma

P. 1393. A prepuce, nearly covered with large masses of warty papillary growths, forming a kind of "cauliflower excrescence."

Penis—Carcinoma

P. 1394. Part of a penis, of which the glans has protruded through a large ulcerated opening on the base of the prepuce. The glans and the exposed surface of the penis, for an inch behind it, are irregularly ulcerated and covered with granulations. The remains of the prepuce are enlarged and very oedematous, but its interior is not exposed.

XXIX

Testicle

P. 1395-1422

Cysts of Vestigial Structures

P. 1395. A testicle, with two small pyriform cysts, attached by short narrow pedicles to the upper part. One hanging from the margin of the globus major represents the end of the Wolffian duct abnormally enlarged.

The other attached to the upper pole of the testicle represents the free end of the Müllerian duct or hydatid of Morgagni.

P. 1396. A testicle, with part of the spermatic cord of a boy. There is a cyst, or "encysted hydrocele," of the cord, rather more than 1 inch in length and about $\frac{1}{2}$ inch from the epididymis. It is formed of a membrane which is smooth and polished internally, and externally is closely connected with the other tissues of the cord, though separable from them. Its origin may probably be referred to a dilatation of an unobliterated portion of the funicular part of the vaginal process of peritoneum.

Cyst of Epididymis

P. 1397. A testicle, to the posterior part of which, but external to the tunica vaginalis, a large cyst, which contained a serous fluid, is attached. The cyst has thick and tough walls, and a smooth, polished, internal surface, beneath which there are some small plates of calcareous matter. At its interior part there is an oval aperture, by which it communicates with a smaller cavity, imbedded in the lower part of the spermatic cord just above the upper end of the epididymis. The testicle is healthy; the tunica vaginalis is thickened, and at the upper part its opposite surfaces are adherent.

> "Another collection (of fluid), we are told, takes place in the cellular membrane of the tunica vaginalis of the cord. A hydatid may form anywhere on the outer part of the testicle, or on the body of the testicle (of

these I have seen several), and may be joined with the true hydrocele, as in this preparation, where one hydrocele formed on the tunica vaginalis testis, which was also distended with water; and hence two distinct cavities of water, one behind the other." Parkinson's *Reminiscences*, p. 91.

P. 1398. A testicle, with several large cysts attached near the head of the epididymis. Both layers of the tunica vaginalis are thickened; its cavity appears to have been the seat of hydrocele. The small pyriform cyst, situated upon the surface of the larger one, probably represents that known as the hydatid of Morgagni. Within the larger cyst is a partition separating it into two unequal portions

P. 1399. The upper half of a testicle, with several small cysts in the epididymis. They are seen through its transparent covering, and one is attached to it by a slender pedicle.

Tunica Vaginalis—Hydrocele

P. 1400. A testicle, with its tunica vaginalis, exhibiting the enlargement and distention of the latter, constituting simple hydrocele. It shows, also, the kind of recess, half-partitioned from the main cavity, which the distended sac forms by the side of the epididymis.

P. 1401. A testicle, with its tunica vaginalis. The sac of the latter was distended in a large hydrocele; it is thickened, opaque, and tough; and its serous surfaces have lost their smoothness. The testicle is situated near the junction of the lower and middle thirds of the posterior wall of the sac, and lies somewhat obliquely. There are two small pedunculated bodies on the upper part of the epididymis.

P. 1402. A testicle, with the sac of the tunica vaginalis, distended in a very large hydrocele. The tunic is but little thickened, and its internal surface is smooth and polished. The testicle is situated in nearly the same part of the sac as in the last specimen, but is placed vertically. There are two small cysts, containing a partly coagulated fluid, in the wall of the sac just above the testicle. The recess by the side of the epididymis is nearly effaced by the great enlargement of the sac.

Tunica Vaginalis—Haematocele

P. 1403. A testicle, with its tunica vaginalis, which was the seat of a hydrocele of moderate size. The condensed haemorrhagic exudate which formed an inner lining to the sac has been dissected from the tunica vaginalis.

P. 1404. A testicle, with its tunica vaginalis, which was the seat of a hydrocele of moderate size. A dissection of the wall of the sac has been made similar to that in the previous specimen.

P. 1405. A testicle, with the greater part of the spermatic cord. The tunica vaginalis is thickened, indurated and consolidated with the surrounding tissues, as in the preceding specimens of haematocele; it is lined with a thick, uneven, granular layer of decolorised blood clot. The testicle is at the lower part of the sac. Above the tunica vaginalis there is a part of the sac of an inguinal hernia.

Tunica Vaginalis—Adhesions

P. 1406. A testicle, with its tunica vaginalis, which was the seat of a large haematocele. The tunica vaginalis is thickened and unnaturally vascular, and lymph is deposited in flakes on its inner surface and on the exterior of the testicle. Around the lower part of the distended sac of the haematocele there is a second sac, with a smooth and free internal surface. The second sac was probably formed by fluid either escaping or being injected between the tunica vaginalis and three layers of haemorrhagic exudate.

P. 1407. A testicle, reduced in size, with complete adhesion of the opposite surfaces of the tunica vaginalis.

P. 1408. A testicle, of unnaturally small size. The lower half of the cavity of the tunica vaginalis is obliterated by the adhesion of its opposite surfaces. The rest of the membrane is thickened. A small pedunculated body is attached to the front and upper part of the epididymis, which is indurated and has an uneven surface.

P. 1409. Sections of a testicle, and of cysts connected with it. The cysts are three in number; the largest is above, and in front of the testicle; the others are behind and below it. The opposite surfaces of the tunica vaginalis are closely adherent over all the front of the testicle, and where they are not adherent they are continuous with the walls of the cysts; so that it is most probable that those which appear alike distinct serous cysts are portions of the tunica vaginalis, which became distended with fluid after the rest of the cavity had been obliterated by adhesion.

Tunica Vaginalis—Calcification

P. 1410. Portion of a tunica vaginalis, thickened, and having two small plates of bone-like substance in it.

Testis—Hypoplasia

P. 1411. A testicle, which is not more than ½ inch in length. Its tubular texture is indi-tinct, and is indicated only by the appearance of cords passing transversely across the substance within the tunica albuginea.

Testis—Calcification

P. 1412. A testicle, reduced in size. Its substance is indurated, and a mass of bone is adherent to the inner surface of the tunica albuginea.

Testis—Hernia

P. 1413. Section of a testicle, of which the upper and anterior part is protruded through an ulcerated opening in the surrounding tissues, in the form of a circular, flat, soft, granulating growth. The surface of the protruded substance is blackened by the action of some caustic. Behind the epididymis, and closely connected with it, is a thickly-walled cyst about 1½ inches in diameter, on the interior of which there are small flakes of inflammatory exudate.

Testis of Bull—Acute Orchitis

P. 1414. "A section of the testicle of a Bull, in a diseased state, showing the tubuli testis"
MS. Catalogue.

They appear to have been unravelled and separated by fluid collected among them.

Testis—Chronic Epididymitis

P. 1415. Section of a testicle, of which the epididymis is enlarged and indurated.

Testis—Gumma

P. 1416. Half of a testicle, injected and sagitally bisected.

In the lower part of the body, which is correspondingly enlarged, there is an irregular lobulated mass of pale yellow, opaque, caseous or necrotic tissue which has been the seat of antecedent fibrosis. The tissue in the vicinity is abnormally fibrous, the fibrotic testicular tissue passing, in places, directly into the necrotic focus. The tissue immediately around the gumma is more vascular than elsewhere. Both the visceral and parietal layers of the tunica vaginalis are thickened from participation in the inflammatory process.

Microscopic examination shows no traces of tubercular histology.

P. 1417. A testicle, in the interior of which there are several opaque masses described as "scrofulous." One of these, at the upper part of the body of the testicle, is cut through, and exhibits at its centre a small necrotic cavity. The sac of the tunica vaginalis is obliterated.

P. 1418. A vertical section of the testicle, injected. In the body of the organ there are three necrotic areas immediately round which the testicular structure is increased in vascularity and is fibrotic.

Microscopic examination of the advancing margin of one of these shows none of the histological marks of tuberculosis, hence it may be concluded that the lesion is syphilitic.

P. 1419. Section of a testicle, described as being enlarged by the deposition of what was regarded as scrofulous matter throughout its substance. The parts intervening between the deposits are dense and fibrous. The testicle has the appearances of syphilitic disease with gummata.

Testis—Teratoma

P. 1420. Section of a testicle, together with some of the surrounding parts, in which there is a large cystic tumour. The greater part of the cavity of the tunica vaginalis, distended by the growth of the tumour, is obliterated by adhesions. A small portion remains free at the front and upper part. Here, also, the section of the tunica albuginea may be seen, and beneath it some of the tubular substance of the testicle, apparently of healthy texture, spread-out in a layer over the upper part of the tumour. The tumour is of regular oval form, smooth on its surface, and measuring about 5 inches and 4 inches in its chief diameters. It is composed almost exclusively of tough-walled cysts, connected into a compact elastic mass by a small quantity of tough, fibrous tissue. The cyst-spaces or tubes were, in places, of considerable length, and smaller tubes opened into the larger ones. Many of them communicated, as was shown by fluid which had been injected into one cyst issuing from others at some distance off.

Microscopically the tumour contains a variety of epithelial and connective tissue structures including cartilage.

P. 1421. Another portion of the same testicle and tumour.

"A large portion of a testicle, which had been converted into a mass of hydatids". Baillie's *Morbid Anatomy*, fasc. viii, pl. viii, fig. 2, p. 182.

Testis—Carcinoma

P. 1422. Parts of a liver, small intestine, omentum, and testicle. In the portion of liver a small spherical mass is seated just beneath its surface. With the intestine are several spheroidal tumours, with shreddy surfaces, attached by broad bases to its peritoneal coat and to the mesentery. The omentum is thickly beset with groups of smaller growths of the same kind. The testis is enlarged and completely replaced by a solid tumour which is divided into rounded lobules by fibrous septa.

Microscopic examination of the growth in the testicle shows it to be a spheroidal-celled carcinoma. The secondary growths exhibit a similar structure.

XXX

Ovary

P. 1423-1431

Ovary—Accessory

P. 1423. A Fallopian tube and ovary. On the Fallopian tube there is a small, oval, pedunculated hydatid of Morgagni; and on the ovary, a small, flattened, round tumour.

Ovary—Follicular Cysts

P. 1424. An enlarged ovary, in the interior of which there are numerous small, oval, smoothly-walled cysts, with distinct thin walls.

P. 1425. An ovary, enlarged by the growth of several simple cysts within it. They are of various size; most of them are oval; others, by their mutual pressure, have lost this form. Some of them have been opened; of these the largest is $1\frac{1}{2}$ inch in its chief diameter, and has tough and smooth walls.

Ovary—Dermoid Cyst

P. 1426. An ovary, in which a cyst, $\frac{1}{2}$ inch in diameter, is filled with a bright yellow fatty substance, and a few small hairs.

P. 1427. The Fallopian tube and ovary of a woman 18 years of age. The ovary contains a cyst 2 inches in diameter, within which there is a second cyst of rather smaller size and with thin walls. This second cyst is filled with a mass of fat and brown hair irregularly mixed, as if they had been rolled together; and at one side of it there is a firm substance like a piece of skin, upon which several fine large hairs and an imperfectly formed tooth are implanted. There is also a small oval cyst attached by a pedicle to the Fallopian tube.

P. 1428. A tough, thick-walled, ovarian cyst, which was filled with fat and hairs. It had been emptied and everted to show that a great part of its inner surface is covered with minute pedunculated wart-like bodies, of which some are nearly isolated, but the greater part are arranged in close-set and elevated groups like warty growths.

P. 1429. An ovarian cyst, of spheroidal form, 6 inches in diameter and filled with a fatty substance, in which numerous fine but stiff, light and curling hairs are interspersed.

Ovary—Cystadenoma

P. 1430. A uterus, with the Fallopian tubes and ovaries. In the place of the right ovary there is a large multilocular cyst. The nature of the contents is not recorded.

P. 1431. A uterus, with the broad ligaments and ovaries, and a portion of a large lobulated tumour connected with the left ovary. The right ovary and other parts are healthy.

XXXI

Fallopian Tubes

P. 1432-1438

Fallopian Tube—Hydatid of Morgagni

P. 1432. The outer extremity of a Fallopian tube, near the fimbriae of which a small, oval, and thin-walled cyst, full of fluid, is attached by a slender pedicle. This cyst, generally known as the hydatid of Morgagni, is believed to represent the closed extremity of Müller's duct.

P. 1433. The ovary and part of the Fallopian tube of a Calf. From the tube hangs a pedunculated body resembling one of the pedunculated cysts extremely common in the neighbourhood of the fimbriae of the tube, haemorrhage having filled its interior.

This specimen was formerly described as an example of vascularisation of a coagulum.

Fallopian Tube—Chronic Salpingitis

P. 1434. Section of a uterus, with one of the Fallopian tubes and broad ligaments. The fimbriated extremity of the Fallopian tube has been turned round, and become closely adherent to the side of the uterus; and, in consequence of the closure of both of its orifices, fluid has collected in it, and distended it into an elongated pyriform sac. The mucous membrane of the tube appears to have given way in some places under the distension; it projects in folds wherever the tube bends suddenly. There is a fibrous tumour in the fundus of the uterus.

Fallopian Tube—Cyst

P. 1435. Portion of a uterus with the right Fallopian tube and ovary. A globular cyst, nearly 2 inches in diameter, is connected with the outer part of the Fallopian tube. The walls of the cyst are tough and laminated; and it has a distinct thin lining membrane, paler than the rest of its walls; its internal surface is for the most part smooth and polished.

P. 1436. The fimbriated extremity of the Fallopian tube of a Sow, laid open to show a fine membranous cyst attached to its inner surface.

P. 1437. The fimbriated extremity of the opposite Fallopian tube of the same animal, with a small pedunculated cyst attached to its outer surface.

Broad Ligament—Cyst

P. 1438. A Fallopian tube and ovary. A thin-walled cyst, measuring about 1 inch in its chief diameter, lies between the layers of the broad ligament in the region of the parovarium, and separate from the ovary. The "hydatid of Morgagni," attached to the longest fimbria of the Fallopian tube, is well developed.

XXXII

Uterus

P. 1439-1460

Uterus—Atresia of External Os

P. 1439. A uterus and vagina, having their cavities laid open from the front. The os uteri is closed; the cavity and walls of the uterus are healthy, and but little enlarged.

P. 1440. The uterus of a Sheep, the horns of which are dilated in consequence of the closure of the upper part of the vagina and of the os uteri.

Uterus—Cyst of Pelvic Peritoneum

P. 1441. The lower and posterior part of a uterus, recently parturient, to the outer surface of which a thinly-walled cyst, about 1½ inches in diameter, is attached.

Uterus of Macacus—Hydatid Cyst

P. 1442. The uterus of a Macacus, surrounded with thin-walled cysts containing hydatids.

Uterus—Interstitial Fibromyoma

P. 1443. A uterus, cut open through the posterior wall, in which is a spherical fibrous tumour, 1 inch in diameter. The tumour has grown in the middle of the wall, and projects slightly both into the cavity of the uterus and externally. It has but a loose connection, by fibro-cellular tissue, with the surrounding uterine substance.

P. 1444. A uterus, in which a fibrous tumour, more than an inch in diameter, has grown in the substance of the fundus, and projects towards the peritoneal cavity.

Another much smaller tumour, formed close by it, projects into the cavity of the uterus, and is flattened by the pressure of its walls. There is a third very small one in the middle of the left wall; and a fourth, of equally small size and disk-shaped, lies just beneath the peritoneum near the right Fallopian tube. This Fallopian tube and the ovary are adherent to the side of the uterus. The cavity of the uterus is enlarged where the tumour projects into it.

P. 1445. A uterus, with the ovaries, part of the vagina. Two nearly spherical, but rough-surfaced, fibrous tumours were formed in the posterior part of the fundus of the uterus, just under the peritoneum, and projected towards the abdominal cavity. Both are calcified. One of them, about $1\frac{1}{2}$ inches in diameter, is partially exposed by dissecting away the peritoneum and part of the uterus covering it; the other is removed, and the cavity in the wall of the uterus in which it lay, and with the interior of which it was loosely connected is exposed.

P. 1446. The tumour shelled-out, from the uterus last described. It is oval, and nearly 2 inches in its greatest diameter, and has undergone patchy calcification.

P. 1447. A uterus, the cavity of which is dilated to a diameter of 6 inches by the growth of a large spherical fibrous tumour within it. The tumour appears to be attached over a wide extent of the wall of the uterus, but it does not protrude from the cavity; its substance is softened and broken down. The walls of the uterus have grown in correspondence with the enlargement of its cavity so that it resembles a uterus at the end of the fourth month of pregnancy. There is a growth like the fibrous tumour, but softer and more degenerate on the outside of the uterus: it is connected with the right broad ligament and ovary.

Uterus—Submucous Fibromyoma

P. 1448. Part of a uterus, to the interior of which, near the upper left angle of the cavity, a small portion of a polypoid fibrous tumour is attached.

P. 1449. Section of a uterus, and of a spherical fibrous tumour, an inch in diameter, in its upper and posterior wall. The tumour presents a somewhat laminated concentric arrangement of its chief fibres. It has grown almost immediately beneath the mucous membrane, which, together with a thin layer of uterine tissue, is reflected over all that part of it which projects into the cavity of the uterus.

P. 1450. A uterus, in the walls of which there are three fibrous tumours. Two of these are situated in the middle of the left wall, and by their mutual pressure have lost the spherical form which they usually assume. The third is situated at the fundus of the uterus, and has grown into its cavity, to the form of which it has in part adapted itself, becoming broad and flat. The cavity of the uterus is enlarged at this part, and its mucous membrane is superficially ulcerated. A small soft mucous polypus is attached by a long pedicle to the wall of the uterus, near the cervix.

P. 1451. A uterus, in the posterior wall of which a fibrous tumour, 9 inches long and 7 inches broad, has been developed. By the growth of the tumour (which reaches from the fundus of the uterus nearly to the outer end of the lower wall of the vagina) the uterus itself has been drawn out to a length of between 7 and 8 inches. Its walls are thin, and exhibit large vessels in them like those of a pregnant uterus; and its cavity, which is lengthened in the same proportion as its wall, is not more than ½ inch in diameter. The tumour, though single, has an aspect as if it were made up of several smaller fibrous tumours; for it consists of many portions of various sizes and shapes compressed together, but not very tightly, and connected by thin and rather loose layers of fibro-cellular tissue, which, on the cut surface, look like partitions subdividing the great mass. The peritoneal surface of the uterus is covered with inflammatory exudate.

P. 1452. Section of a uterus and of an enormous fibrous tumour growing from the posterior part of its cervix. The tumour is of an elongated oval form, 12 inches in length and 5 inches in thickness. At its lower end there are several small smooth-walled cavities and at the upper part the boundary between the tumour and that part of the cervix uteri which was expanded over it as it grew, can be distinctly traced. At this part the substance of the uterus is much increased in thickness and vascularity, and can be traced in a thin layer over a great portion of the posterior surface of the tumour. The anterior wall of the uterus, the upper half of its posterior wall, and its cavity are of natural size: the mucous membrane of the latter is superficially ulcerated. The anterior surface of the tumour is rough, as if from superficial ulceration and its lower end, which must have protruded far from the vagina, is in a similar state. The bladder and rectum are connected, as usual, with the uterus; but there is a large quantity of inflammatory exudate in the pouch between the uterus and the rectum.

Uterus—Calcified Fibromyoma

P. 1453. Sections of a large brownish coral-like mass, oval in form and about 5 inches in its chief diameter, and very compact and heavy, which was found in a grave-yard and sent to Mr Hunter as a urinary calculus. It was probably formed in a fibrous tumour of the uterus. It yielded, on analysis, 18·644 per cent of animal matter, consisting of gelatine with a small proportion of albumen. Its other chief constituents were found to be phosphate and carbonate of lime, the proportion of the carbonate being apparently greater than it is in human bone.

Uterus—Polypoid Adenoma

P. 1454. A uterus, with a firm fibrous tumour attached to a considerable portion of the length of its posterior wall. The tumour has grown into the uterine cavity, to the elongated and flattened form of which it has adapted itself.

P. 1455. "The uterus of a woman, showing a polypus that grew from the inside of the uterus, and, as it grew, was conducted into the vagina, where it had more room, and there swelled." *MS. Catalogue.*

The pedicle by which the polypus is attached to the middle of the left wall of the uterus is 1 inch long and $\frac{1}{4}$ inch thick; its tissue may be traced, by its white fibrous bands, into the substance of the uterus. The mucous membrane of the uterus is continued over the polypus, but is in parts excoriated. The vagina is enlarged. The broad ligaments and other parts near the uterus are healthy.

Microscopic examination proves the growth to be an adenoma which has arisen in the uterine mucosa.

Uterus—Endocervical Carcinoma

P. 1456. A uterus, of which the cervix, together with all that part of the vagina which was attached to it, has been destroyed by cancerous ulceration. The uterus itself is enlarged near the ulcerated part, but the ovaries, Fallopian tubes, and broad ligaments are healthy.

P. 1457. A uterus, with the ovaries, and other adjacent parts, exposed from behind. The cervix, and the lower half of the body of the uterus, are involved in a large, soft, ulcerated growth. The uterus is enlarged, and at the upper part of its cavity the sinuous and slightly elevated border of the ulceration may be discerned. The ovaries, bladder, and other adjacent parts are healthy.

Uterus—Cervical Carcinoma

P. 1458. A uterus, with the ovaries, broad ligaments, vagina, and other adjacent parts, exposed from behind. A broad flat growth is attached to all the right and posterior part of the cervix uteri. The growth has a sharp, projecting and overhanging, irregularly sinuous border; it measures about 4 inches from side to side, 2½ inches from before backwards, and nearly 1 inch in thickness at its middle and thickest part. A ligature was tied round a part of its base, its substance having been first broken through; but the part tied does not appear altered in its texture. The uterus is large; its cavity is dilated; and on its left wall there is a small flat growth, ½ inch in diameter, like a mucous polypus. The ovaries are both adherent to the sides of the uterus and the broad ligaments are thickened.

P. 1459. A uterus, with the ovaries, vagina, and other parts, exposed from behind. The prominent part of the cervix uteri, and all the adjacent portion of the vagina, are deeply ulcerated. The ulcer has an uneven shreddy base; and on the left side of the vagina there is a long piece of tissue, which has apparently sloughed from it. The border of the ulcer is not well defined or elevated; the tissues beneath it are thickened and indurated; its nature is doubtful, but most probably it arose as a cervical cancer. The body of the uterus appears healthy: the ovaries are adherent to it. The mucous membrane of the bladder is superficially ulcerated; and the middle of its posterior wall is perforated by the ulceration extending from the upper part of the vagina.

P. 1460. Parts of a uterus, urinary bladder, and vagina. The posterior wall of the bladder, the lower half of the uterus and all the upper part of the vagina, have been destroyed by carcinomatous ulceration. Their remains are ulcerated and shreddy, and have numerous large sloughing and nearly detached portions of ulcerated tissues hanging from them. The enlarged upper half of the body of the uterus remains, and a quill is placed in its cavity. This, and the remains of the urinary bladder and of the external part of the vagina, serve to mark the relative positions of the organs; their forms are hardly discernible.

XXXIII

Female External Genitalia

P. 1461-1468

Labia Minora—Hypertrophy

P. 1461. Two nymphae, enlarged and indurated. Their surface is like that of healthy skin, but lobed and wrinkled.

P. 1462. Nymphae, enlarged and indurated.

P. 1463. Nymphae, enlarged and indurated. The right nympha is more than four times as large as the left: it is of an oval form, measuring 3 inches in length and 1½, inches in thickness. A section shows that it is composed of a uniform pale dense, skin-like texture; its surface is coarsely wrinkled and it is covered with thin cuticle.

"Removed from a woman at St. George's Hospital by Mr. Gunning."

MS. Catalogue.

Clitoris—Hypertrophy

P. 1464. Parts of the genital organs of a woman, enormously enlarged. The parts preserved consist of four distinct portions, two at the sides and two in the middle between them. The two lateral portions are probably the labia, each of which is 6 inches in length and 2 or 3 inches in width. Their surfaces are dark, deeply lobed, and wrinkled; they appear to be composed of firm skin-like tissue, and are covered with a thin epidermis. These two portions are united above by a narrow isthmus, from the front of which are suspended the two middle portions, which may be considered as the enlarged clitoris and its prepuce. They form two irregularly-rounded masses, attached above by narrow pedicles, and presenting the same external characters as the enlarged labia. The lowest of the two, which is also rather the largest, measures 4½ inches in one, and 3 inches in another diameter. The section of one of them shows that it is composed of a compact, pale, yellowish-white tissue, like that of tough skin, uniform, and constituting the whole of the enlargement, not isolated like a tumour imbedded in other tissue.

Vagina—Acute Inflammation

P. 1465. A uterus and part of a vagina. The substance of the uterus is extremely dense and hard, but its form is unaltered. In the upper part of the vagina there are numerous small ulcers. Some of them are oval or round, others irregular in their outlines; they all have sharp, abrupt, but not elevated, borders, and their bases are smooth, and do not appear to be indurated. None of them penetrates further than the submucous tissue.

Vulva—Acute Inflammation

P. 1466. The labia pudendi of an infant, enlarged, swollen, and superficially ulcerated on their anterior surfaces.

Vagina—Carcinoma invading Bladder

P. 1467. A uterus, vagina, and urinary bladder. A large irregular opening has been formed, probably by cancerous ulceration of the adjacent walls of the vagina and bladder, and around the aperture there are sloughs. There are also remains of adhesions over the whole peritoneal surface of the uterus and ovaries.

P. 1468. Parts of a vagina and bladder. A large, oval, circumscribed, and probably cancerous ulcer, extends around the whole circumference of the vagina, and presents a rough sloughing surface. At one part the ulceration has extended through the posterior wall of the bladder.

XXXIV

Gestation

P. 1469-1478

Gestation—Retained Foetus

P. 1469. A portion of the skin and subjacent muscles of a Calf which was retained in the uterus two years beyond the ordinary period of gestation. All the tissues appear fresh and healthy.

P. 1470. One of the feet of the same Calf, the tissues of which are equally sound.

P. 1471. Portion of the horn of the uterus of a Sheep, containing the head and one of the feet of a lamb, which remained in the uterus beyond the ordinary period of gestation and became adherent to the surrounding uterine wall.

P. 1472. Section of the wall of the same uterus, exhibiting strong and well-organised adhesions between its inner surface and the skin of the neck of the lamb.

P. 1473. "A part of the uterus of a Sow, in which the young pig had died and become rotten, by which the mother also died." *MS. Catalogue.*

Gestation—Ruptured Uterus

P. 1474. "The lower half of a gravid uterus, which had arrived at the full period of gestation. On the patient accidentally falling, her labour pains came on; and before she could be delivered the uterus burst, and the child's arm made its escape out of the uterus, but did not pierce the peritoneum at the part where it was reflected over the bladder, uterus, and inside of the pelvis" (*MS. Catalogue*). There are two distinct lacerations of the uterus, one on the left lateral wall, 4 inches long and nearly vertical, the other on the right and posterior wall, near the cervix, of the same length, but transverse. Parts of the bladder and rectum are preserved *in situ*, and are healthy; the aperture in the latter appears to have been made after death.

P. 1475. The right half of a gravid uterus, which was lacerated in parturition at the full period of gestation. There are two lacerated apertures in the wall of the uterus, one at the side, nearly 3 inches in diameter, the other posteriorly, about 1 inch in diameter, and both low down in the cervix. Blood is effused on the borders and outer surfaces of the lacerations, and there is lymph on the adjacent part of the peritoneum. The os uteri is widely dilated; the bladder is contracted and healthy.

P. 1476. Part of a uterus, which was lacerated in parturition at the full time of gestation. The laceration is through the anterior wall, $1\frac{1}{2}$ inches from the os uteri, and just above the part at which the peritoneum is reflected from the posterior wall of the bladder. It is transverse, and between 3 and 4 inches in length; its margins are smoothly rounded, as if it had existed for some days before the patient's death. The walls of the uterus, also, are contracted, and its great veins are empty and collapsed.

The child was found in the cavity of the abdomen.

Gestation—Vesico-vaginal Fistula

P. 1477. A uterus, with the vagina, external organs, and bladder. There is an irregular aperture of communication, nearly 2 inches in diameter, between the vagina and bladder, through their adjacent walls. This aperture was the consequence of ulceration following extensive injury of the parts in parturition. In the subsequent contraction and adhesion of the parts, the walls of the bladder, which formed the borders of the aperture, appear to have been drawn backwards, so that the ureters (indicated by bristles) now open on the upper wall of the vagina, one on each side of the laceration. The remaining part of the cavity of the bladder is healthy, but closely contracted. The interior of the vagina is ulcerated and incrusted with calcareous matter. The uterus, ovaries, and external organs are healthy.

Gestation—Extrauterine Pregnancy

P. 1478. An ovum with the decidua reflexa and a large mass of coagulum attached thereto; the cavity of the chorion is laid open, exposing a pyriform membranous sac with a pedunculated attachment to the inner surface of the chorion.

XXXV

Breast

P. 1479-1485

Breast—Chronic Mastitis

P. 1479. Section of a tumour of the breast. It consists of a pale firm substance, with numerous small thinly and smoothly walled cysts in its interior, and with one cyst of the same kind, 1 inch in diameter, lying near it.

Breast—Duct Papilloma

P. 1480. Part of a breast, in which, just below the nipple, there are two cysts separated by a thin membranous partition. They are of oval form, and appear to have been about $1\frac{1}{2}$ inches in diameter; they have distinct, thin, membranous walls with polished internal surfaces. At the part nearest to the nipple there protrudes into the cavity of each of these cysts a coarsely-lobed and granular growth of soft substance, like a mass of granulations. Both the growths have broad bases of attachment to the walls of the cyst; and their free surfaces are convex and deeply fissured.

Microscopic examination of the mammary tissue in the left-hand edge of the specimen shows it to be the seat of chronic inflammation.

P. 1481. Sections of part of a breast and of a tumour immediately below the nipple. The section has been made through the axis of the nipple, and bristles are placed in the lactiferous ducts. The tumour is about $\frac{3}{4}$ inch in diameter and completely fills a smooth-walled cyst formed by a dilated lactiferous duct. The texture of the surrounding mammary gland appears quite healthy.

"The tumour was very movable. The first sign of the disease was occasional bleeding from the nipple; some time after this the tumour was first felt."

Everard Home's *Observations on Cancer* (London, 1805), Case v, p. 45.

Breast—Fibroadenoma

P. 1482. Sections of a small round tumour in a mammary gland. The surface of the tumour is distinctly defined and loosely connected with the adjacent textures. Two thirds of it are composed of a pale compact tissue, with ill-defined fibrous septa. The other third is made up of small leaf-like pedunculated growths filling cysts, from which some have been everted, and in which the others are so tightly packed that they form a nearly solid and uniform substance resembling that of the rest of the tumour. The exterior of the tumour is invested by a thin fibro-cellular capsule; and the adjacent tissues appear healthy.

Breast—Carcinoma

P. 1483. Section of a breast, with a hard and large cancerous tumour in the substance of the mammary gland.

Microscopic examination shows the tumour to be a spheroidal-celled carcinoma of the soft or encephaloid type.

Breast—Cancer

P. 1484. Section of a breast, and of a small hard cancerous tumour immediately beneath the deeply retracted nipple. Several round cords, supposed to be thickened lactiferous ducts, have been dissected out on one side of the tumour. They contained a whitish fluid.

Breast—Carcinosarcoma

P. 1485. Section of a breast, with a tumour, in which "the cancerous suppuration had taken place". The substance of the tumour, which is situated just below the deeply retracted nipple, is soft and broken, from caseous degeneration, so as to present many cavities and slight fissures, with a generally granular and grumous aspect. In its form and intimate connection with the adjacent tissues it is like a hard cancer.

Microscopic examination shows that the tumour is a spheroidal celled carcinoma, in the deeper part of which the stroma is sarcomatous and of the spindle-celled form.

XXXVI

Skin

P. 1486-1527

Skin of Dace—Mycosis

P. 1486. A Dace, the surface of the body of which is nearly covered with a thick felted mycelium.

Tail of Horse—Actinomycosis

P. 1487. A section of the tail of a Horse, and of a large, lobulated, soft, pale-brown and black mass imbedded deeply in its substance, and projecting in lobes and nodules on its surface.

Skin of Leg—Elephantiasis

P. 1488. A transverse section of a leg affected with elephantiasis. The integuments are nearly 1 inch in thickness, and are very firm. On their cut surface they appear composed of a tough, white, fibro-cellular tissue, forming thick bands and laminae, which enclose areolae resembling those of the deep layers of coarse skin, and filled with masses of a deep yellow, transparent, fatty substance. The surface of the skin is dark, hard, and very coarsely wrinkled and granular. The cuticle is not of unusual thickness. The bones and muscles appear healthy.

P. 1489. Another section of the same leg, taken from its lower part, and including the lower end of the tibia, the articular surface of which is healthy.

P. 1490. The foot and ankle of the same person. The whole integument is similarly diseased, but its surface is paler and smoother than in the preceding specimens.

Skin of Foot—Corn

P. 1491. Part of a foot, from the sole of which a portion of the epidermis has been reflected in the region of a corn. The depression in the cutis, in which the corn was embedded, is shown. The great toe is turned outward and part of its nail rests in a depression on the back of the second toe.

P. 1492. Section of the sole of a foot, with a very small corn formed by a circumscribed mass of hard epidermis, which is imbedded in the surrounding healthy epidermis, but exhibits a line of separation all round it.

Skin of Toe—Corn

P. 1493. A toe, on the back of which there is a corn over the first articulation, and beneath the corn a small bursa. The corn is reflected with the cuticle, and the bursa is shown by a lateral section.

P. 1494. The cuticle of a toe, separated with a corn, and inverted. A short dark process of horn-like substance projects outwards from the middle of the corn.

P. 1495. A toe, from which part of the cuticle has been reflected, with a corn which was situated on the dorsal surface of its first articulation. The corn consists of a circular slight thickening and induration of the cuticle, in the centre of which there is a small and very hard mass, separated from the rest by a narrow groove, and projecting inwards. The depression of the skin, in which the corn was imbedded, is shown and presents an exact mould of its form.

P. 1496. The toe of a Fowl, with a corn upon its margin.

Finger Nail—Deformity

P. 1497. The end of a finger, in which, coincidently with an abnormal growth of the nail, the matrix of the nail has become narrow and unnaturally convex, with a ridge passing along the middle of it.

Skin—Acne Rosacea

P. 1498. A lobulated growth of skin from the nose. The sebaceous glands are enlarged and their orifices prominent.

Skin—Molluscum Contagiosum

P. 1499. A tumour, $\frac{3}{4}$ inch in diameter, from an eyelid.

Microscopic sections exhibit the lobulated epithelial structure of molluscum contagiosum with the characteristic molluscous bodies.

P. 1500. A tumour, removed from an upper eyelid. The greater part of its contents, which consisted of a substance like soft cheese, have been removed. Its internal surface is remarkably sacculated and reticular with minute ridges or folds.

Microscopic sections of the wall demonstrated that the lesion is a much excavated example of molluscum contagiosum.

The following case notes most probably refer to this specimen.

"An Encysted Tumour of Oil in the Orbit

"A young gentleman had a small tumour appearing in the upper and under part of the orbit, which at first was no larger than a pea, but increased in size, becoming more oval, and extending towards the nose. It was situated between the bony orbit and the upper eyelid, which it pressed down, keeping that eye half shut, but unattended by pain, except when engaged in reading by candle-light there was an uneasiness and throbbing in the eye.

"The tumour evidently contained a fluid, was not attached firmly to the inner side of the orbit, but seemed formed in the cellular membrane; upon opening it, the contents were pure oil, perfectly clear and sweet, burnt with a very clear light, and did not mix with aqueous fluids, and, when exposed to cold, became as solid as the human fat.

"This appears, therefore, an encysted fatty tumour, differing from fatty tumours in general; there being no cellular membrane formed, only the oil deposited."

MS. Notes. *Cases in Medicine and Cases in Surgery* by Sir Everard Home.

Skin—Sebaceous Cyst

P. 1501. A cyst, removed from below the chin of a young woman at St. George's Hospital. It has been opened and emptied.

P. 1502. A small globular, cutaneous cyst, with the skin covering its anterior half.

P. 1503. A small, oval, thick-walled cyst, from a scalp. Half of it has been emptied; the other half is filled with a dense and regularly laminated, sebaceous, or almost horny substance.

P. 1504. A Lark with a small oval cyst, containing a soft fatty substance, on the upper part of each of its wings.

Skin—Papilloma

P. 1505. A large spheroidal, warty, cutaneous growth, deeply lobed and nodulated, removed from the external labium of a woman. It was fixed by a very small pedicle.

P. 1506. Sections of a warty cutaneous growth, displaying a pale compact base, composed of tissue like that of the denser layers of the skin, and bearing on its surface bunches and groups of warty nodules fixed by narrow pedicles.

P. 1507. Sections of a wart-like growth from the skin. The greater part of it is composed of tough tissue like that of skin; it is attached by a narrow pedicle, and covered with long and slender thorn-like, sharp processes of the same tissue, which radiate in all directions, and are invested with a thick dark cuticle.

P. 1508. Section of a papilloma, probably not from the human subject. It measures 2 inches in diameter and had, apparently, a broad base. The spines covering its surface are for the most part grouped, and they taper from broad and, in some instances, suddenly expanded bases.

P. 1509. A section of a papilloma from the leg of a Dog. Some of the spines on its surface are ½ inch in length, and many of them are united in groups.

Skin—Angioma

P. 1510. Sections of a tumour removed from a woman's back. The tumour is lens-shaped, 1½ inches in diameter, and constricted at its base. It is partly covered by skin, the papillae of which are enlarged. Its cut surfaces present the appearance of a soft, uniform, brain-like substance.

Histologically the tumour has the structure of a capillary angioma, possibly of the hyperplastic kind since no trace of blood is present in the spaces. The vessels are mostly collapsed, and this together with the presence of intervening connective tissue conveys the impression that the growth is more solid than it really is.

P. 1511. "A tumour from under the lower jaw" (*MS. Catalogue*), an example of erectile tumour. The cut surface of the tumour displays a close network of fine smooth shining bands and cords, almost exactly like that of the corpus cavernosum penis, only less regular in their arrangement. The exterior of the tumour is uneven, and has no definite character.

P. 1512. A small subcutaneous tumour of spongy texture.

Skin—Pedunculated Fibroma

P. 1513. A small pendulous tumour from the skin of the abdomen.

P. 1514. A cutaneous fibrous growth, attached by a very slender pedicle to the skin of a man's arm.

P. 1515. A pedunculated fibrous tumour of the skin.

P. 1516. A deeply lobed fibrous tumour, which was attached by a narrow pedicle to a breast.

Skin—Lipoma

P. 1517. A large fatty tumour, divided into several lobes, all smooth on their surface, and all closely packed together so as to form a regularly oval mass.

It was removed from above the breast of a young woman.

P. 1518. Section of a large fatty tumour removed from the back. It appears firmer than the preceding, and consists, like it, of many closely packed lobes.

Skin—Pedunculated Lipoma

P. 1519. Section of a fatty tumour and of the skin by which it is thinly covered. It formed a pendulous broad-based tumour on the groin.

P. 1520. A small tripartite and lobulated fatty tumour, covered by skin and attached by a narrow pedicle to the skin of the axilla, from which it was removed.

P. 1521. The head and neck of a Goose, with a large tumour attached near the angle of the lower jaw, on the right side.

Microscopic examination by S. G. Shattock showed the growth to be composed of adipose tissue.

Skin—Lipomyxoma

P. 1522. Portion of a large tumour, removed from the thigh: it was supposed to be an aneurysm. Its cut surface is white, lobed, and obscurely fibrous in structure; and at the same time, in the smallness and leaf-like form of many of its lobules, it has some likeness to the solid and compact growths within the sero-cystic tumours.

Histologically the growth has the characters of a myxoma. The more opaque areas with which the growth is strewn are the more rapidly growing and more cellular foci. Many of the cells in these areas are in various degrees distended with fat.

Skin—Angiosarcoma

P. 1523. Sections of a tumour removed from the back. There is an appearance in it of a cavity containing a clot of blood. The entire growth was, probably, a fatty tumour containing more than the usual proportion of connective tissue.

Histologically the tumour may be classed as an angio-fibro-sarcoma, the fibrous septa between the spaces being somewhat abundantly provided with cells.

Skin—Alveolar Sarcoma

P. 1524. "A tumour of the cellular membrane" (*MS. Catalogue*). It has an oval form, and is 1½ inches in diameter; its surface is nearly smooth; it is composed of a firm, nearly white substance, with short fibres undulating through it, in the form of networks and incomplete partitions. A portion of skin is attached to its surface.

Histologically the tumour has the structure of an alveolar sarcoma.

Skin—Sarcoma

P. 1525. Part of a large, soft, vascular tumour formed beneath the integuments of the brow and pressing the eye downwards. The posterior part of the tumour is necrotic and fragmented.

Skin of Cheek—Columnar-cell Carcinoma

P. 1526. A cyst containing several lobulated growths, which was removed from the cheek of an aged woman at St. George's Hospital. The cyst is globular in form and about 3 inches in diameter; its walls are a line in thickness, and formed of tough and, apparently, fibro-cellular tissue; externally it appears to have been closely connected with the adjacent parts; internally its walls, where no growths are attached to them, are uneven and superficially tuberculated, but have a polished surface. The lobulated growths are numerous, and of various sizes; some are attached by broad bases; others, including the largest, have long and slender cord-like pedicles; they are all composed of a pale soft substance.

Microscopically the tumour is a carcinoma, the cells in some of the alveoli being columnar but in most spheroidal. The surface has a papillary structure. The cyst shows no epithelial lining.

Skin—Melanoma

P. 1527. Sections of a tumour removed from the back. It has a lobed and nodulated surface, a narrow base of attachment, and an irregular outline; its general form is lens-shaped. It is composed of a smooth and close-textured brain-like substance which is streaked and shaded by lines and dots of melanotic matter, dividing it into irregular lobes. The tumour appears to have grown in the cutis, and to have protruded through its surface.

Histologically the tumour is an alveolar sarcoma. In certain of the spaces some of the cells are elongated so as to appear spindle-shaped, in others they are polymorphous or spheroidal.

XXXVII

Teratology

P. 1528-1545

Anterior Dichotomy of Skull

P. 1528. A skull, in which dichotomy extends as far back as the basi-occipital. In this skull there are two supra-occipital and a median external auditory meatus; the periotic bones of the median ear are double; both faces are distinct and well-formed.

Thoracopagy

P. 1529. A human female thoracopagous twin-birth about 12 inches long (about the end of the sixth month of gestation).

P. 1530. Human female thoracopagous twins about full time of gestation. The deficiency of the anterior abdominal wall above the umbilicus is very considerable, the viscera being covered by membrane only.

Anterior Dichotomy of Spine

P. 1531. A foetal Pig with dichotomy of the whole spinal column and four distinct anterior extremities. The thoracic and abdominal cavities have been opened to show the heart and main arterial trunks. The heart consists of two left hearts united; there are no pulmonary hearts. Each ventricle gives off an aorta which crosses to the opposite side, so that the heart of the left side supplied the right inferior extremities, and that on the right the left. The carotids, two in number, are given off together from the arch of the right aorta. The trachea is single. The lungs are very small and imperfect; they probably consist of a single pair. The vessels of the right heart are injected. The palate is deeply cleft.

Anterior and Posterior Dichotomy of Spine

P. 1532. A foetal Pig with posterior dichotomy (schizorhachis) combined with anterior dichotomy of the lower jaw and tongue. There is a considerable cleft in the palate connected with the median rami of the jaw by a fold of mucous membrane.

P. 1533. A female human foetus with anterio-posterior dichotomy (schizorhachis combined with anterior dichotomy), producing considerable doubling of the face. There are four eyes, but the inner ones are contained in a single orbit. The axes diverge both in front and behind, from a point in the basiphenoid bone.

Parasitic Hind Limbs

P. 1534. The bones of the pelvis and inferior extremities of a Sheep, with the pelvic bones and posterior extremities of an imperfect twin similarly attached. In this preparation the parasitic limbs are fused as far as the metatarsus.

Craniopagus Skull

P. 1535. The skull of a Bengalee child, with a second imperfect skull attached to the anterior fontanelle, which is widely extended. The base of the second skull is much contracted, the foramen magnum being almost closed. All the bones of the face are much smaller than those of the supporting skull. The faces are turned in almost opposite directions, so that the left frontal bones of one skull articulate by suture with the right parietals of the other, and vice versa.

The history of this skull and an account of the child are given in the *Phil. Trans.*, vol. lxxx, p. 296, from which the following particulars are extracted:

The child was a male; it was more than four years old at the time of its death, which was caused by the bite of a cobra. It was very emaciated, a fact attributed to the parents having used it as a show, always keeping it covered up, except when payment was made for its exhibition. The woman who acted as midwife was terrified at the appearance of the additional head, and tried to destroy the child by throwing it on the fire; it was rescued after one eye and ear were considerably burnt.

There was no trunk to the second head; but it was surmounted by a short neck terminating in a rounded tumour, which is stated by one observer to have been quite soft at the age of two, and by another to have been quite hard and cartilaginous at the age of four. Its external ears were represented by mere folds of skin, and there was no auditory meatus. The normal face and head were not malformed. The brains were distinct, each invested in its own membranes; the dura mater of each adhered to that of the other at the point of contact. The chief supply of blood to the upper head was by a number of vessels passing from the membranes of one brain to that of the other. The movements of the features of the upper head appear to have been purely reflex, and by no means to have been controlled by the feelings or desires of the child. The movements of the eyes of the accessory head did not correspond with those of the child, and the eyelids were usually open, even during sleep.

Defective Jaws

P. 1536. The skull of a new-born Calf with a cleft between the maxillary bones, and defect of the palate-bones and pterygoid processes. The whole skull is shorter and broader than the normal type. There is considerable shortening and deformity of the mandible.

Defective Facial Axis

P. 1537. The head of an hydrocephalic foetal Pig, with deficiency of the facial axis and fusion of the eyes into a single globe. The vertex is occupied by a very large cyst.

Nanocephaly

P. 1538. A nanocephalous foetal Puppy with no eye or orbit; it is otherwise similar to the last.

Defective Abdominal Wall

P. 1539. A female Human foetus, probably at about the seventh month, with deficiency of the abdominal parietes. There is a large visceral tumour covered by transparent membranous walls. The tumour contains the liver and the greater part of the intestinal canal. There is also arrested development of the forearms. The hands are placed at right angles to the limb. There are but three fingers on the left and four on the right hand. The left leg is shortened. There is curvature of the spine and spina bifida in the lumbar region.

P. 1540. A female anencephalous human foetus with ectopia of the viscera. The sac containing the protruding intestine has been laid open; a large portion of the placenta is seen adhering to the sac.

These two preparations are interesting as they show a frequent complication of these conditions. On *a priori* grounds it would appear likely that the placenta would become adherent to an open membranous surface like the peritoneum, arachnoid, or pia mater, or to the edge of an ulcerating surface. It is not a little remarkable that so many have considered such adhesion as the cause of the malformation, as it is far more probably an effect, or an accidental complication extremely prone to occur.

Extroversion

P. 1541. The bones of a foetal Calf with strong backward curvature of the spine and doubling back of the ribs. The pelvic bones are closely approximated, and several of the ribs are fused together. There has evidently been complete ectopia of the viscera. No anterior extremities existed. This condition is nearly identical with that described by Gurlt under the term *Schistozomus reflexus*, except in the absence of the anterior extremities.

Arrest of Pelvis and Hind Limbs

P. 1542. A foetal Pig, in which the development of the posterior limbs and pelvic girdle has been considerably arrested. The femora are deficient from atrophy, so that the limbs are only attached to the trunk by skin and ligamentous tissue. The abdomen has been opened below to show the extremely diminutive condition of the pelvis and pelvic viscera. The hind legs terminate in blunt points, and each has a very imperfect hoof.

Free Martin

P. 1543. The generative organs of a very young Calf, described as "Palmer's Free

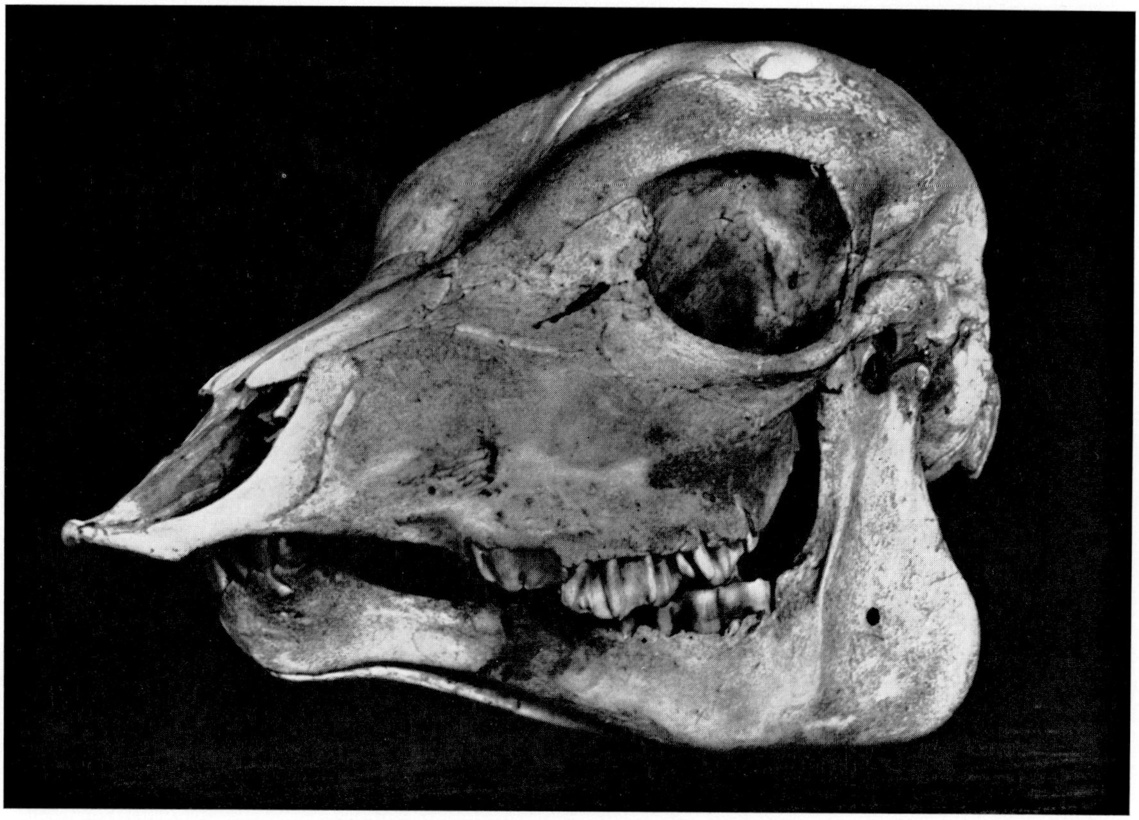

FIG. 21
The skull of a new-born calf showing deformity of the jaws. Specimen 1536.

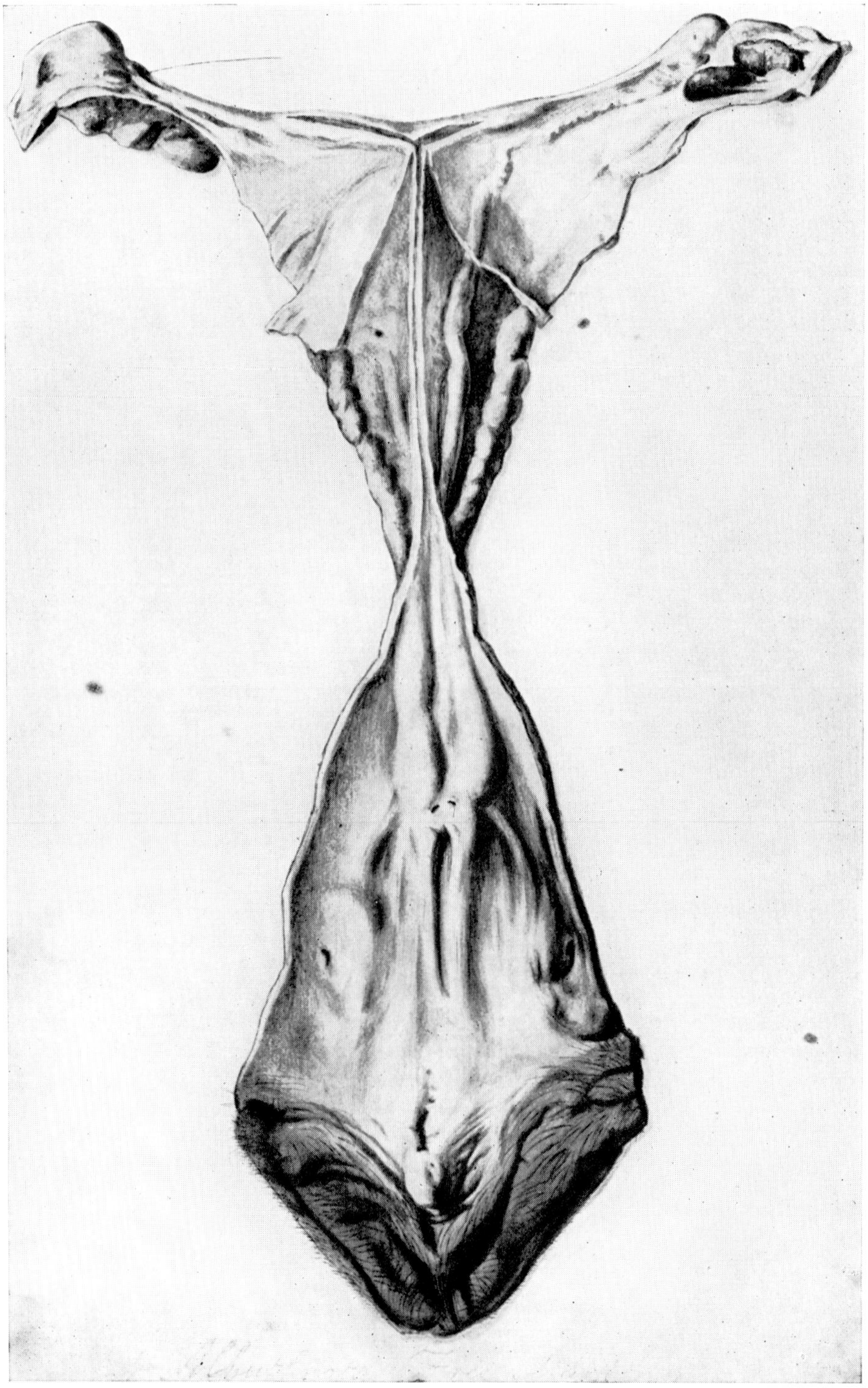

FIG. 22

The organs of reproduction from "Mr Arbuthnot's Free Martin." From a drawing probably by Jan van Rymsdyk. Specimen 1544. Drawing Book C, p. 842.

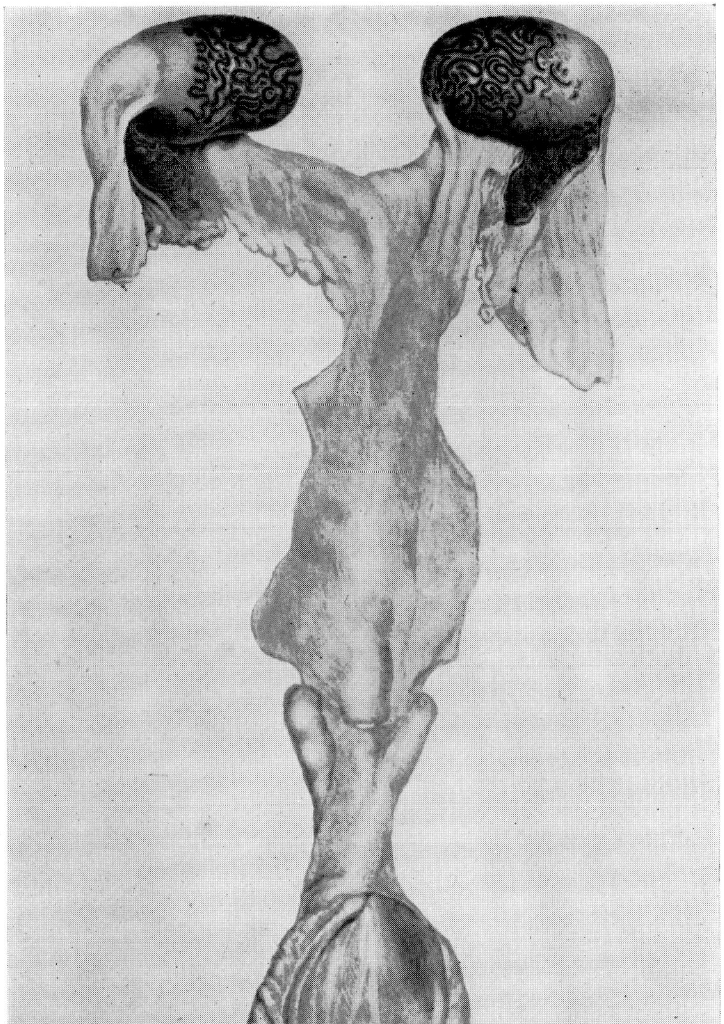

FIG. 23
A drawing, probably by William Bell, of the generative organs of Mr Wright's Free Martin. Specimen 1546. Drawing Book C, p. 841.

Fig. 24
Mr Wright's Free Martin: from an original drawing by Sawrey Gilpin (1733–1807). Drawing Book C, p. 841.

Martin." They exhibit a highly androgynous condition. The large urethral bulb is laid open from above to show the orifices of the urethral and sexual ducts.

Just within the orifice of the bulb which forms the external opening, and closely resembling the vagina of the female, a slight depression is seen, and immediately beneath this, in the anterior wall of the bulb, is an exceedingly diminutive penis; it is retracted permanently in sigmoid curves. The neck of the bladder opens into the sinus of the bulb near its upper extremity, and just above it are the openings of the vasa deferentia; the bulb is prolonged slightly above these, and ends in a thick solid cord, the sexual cord of the embryo. This terminates above in two solid cornua, $1\frac{1}{2}$ inches in length; a pair of rudimentary testes are suspended in a peritoneal fold at their extremities. The epididymis is much larger than the gland, and this has given rise to the opinion that both ovaries and testes exist: sections have been made through these parts. The microscope shows that the testes consist of dense fibrous tissue with cells; the character is not that of an ovary, but closely resembles the testes usually found in these cases. The vasa deferentia are incorporated with the sexual cord above, but are well formed below, as are also the vesiculae seminales. *Works*, vol. iv, p. 39.

P. 1544. The generative organs of an androgynous Bull. The animal was described as "Mr. Arbuthnot's Free Martin."

This preparation is essentially similar to the preceding, except that it is much larger, and the genital cord is perforated by Müller's ducts; these have been laid open through their entire length; they are seen to terminate blindly near the extremity of the cornua. The vesiculae seminales are normal, and the vasa deferentia are developed; they pass up in front of the genital cord for a short distance, and then extend as convoluted tubes along its outer borders, to the extremity of the cornua; they terminate in the epididymis. The testes lie close to the extremity of the cornua; they are very small in proportion to the epididymis, and hence they might easily be mistaken for ovaries. The openings of Cowper's ducts and of the vasa deferentia are marked with purple, and the urethra with a green glass rod. *Works*, vol. iv, p. 42.

P. 1545. A similar but much smaller preparation, described as "Mr. Wells's Free Martin."

The vesiculae and their ducts, as well as the terminal portions of the vasa deferentia, are normal. Cowper's glands are present, and there is a rudimentary penis retracted in sigmoid curves. The sexual cord is solid, as are likewise its cornua. There is only a trace of an epididymis on the left side; but the cornua of the sexual cord end in a large thick-walled cyst: the vas deferens on this side is deficient in several places. On the right side there is no trace of a vas or epididymis, and a small oval body, more like an ovary than a testis, exists in the place of a testis; it consists of tissue, however, which, on microscopic examination, appears identical with that of which the imperfect testes of these malformations usually consists; there is no reason to regard it as otherwise than an imperfect testicle. There is much fat in the neighbourhood of these rudimentary sexual bodies. The following history is given by Hunter in his *Animal Œconomy*:

"It was between three and four years old when killed; and had never been observed to show any signs of desire for the male, although it went constantly with one. It looked more like a heifer than Free Martins usually do. The teats and udder were small, compared to those of a heifer."

Works, vol. iv, p. 43.

P. 1546. The reproductive system of "Mr. Wright's Free Martin, five years old."

Hunter's description of this animal is as follows:

"This animal had more the appearance and general character of the ox, or spayed heifer, than of either the bull or cow. The vagina terminated in a blind end, a little way beyond the opening of the urethra, from which the vagina and uterus were impervious. The uterus, at its extreme part, divided into two horns. At the termination of the horns were placed the testicles, instead of the ovaria, as is the case in the female. The reasons why I call these bodies testicles are the following. First, they were about twenty times larger than the ovaria of the cow, and nearly the size of the testicles of the bull, or rather of those of the ridgil, the bull whose testicles never come down. Secondly, the spermatic arteries were similar to those of the bull, especially of the ridgil. Thirdly, the cremaster muscle passed up from the rings of the abdominal muscles to the testicles, as it does in the ridgil.*

"There were the two bags placed behind the bladder and the uterus. Their ducts opened into the vagina, a very little way beyond the opening of the urethra; but there was nothing similar to the vasa deferentia.

"As the external parts had more of the cow than the bull, the clitoris, which may be reckoned an external part, was also similar to that of the cow, not at all in a middle state, between the penis of the bull and the clitoris of the cow, as I have described in the hermaphrodite horse. There were four teats: the glandular part of the udder was but small.

* "Although I call these bodies testicles, for the reason given, yet they were only imitations of them, for when cut into they had nothing of the structure of the testicle; not being similar to anything in Nature, they had more the appearance of disease. From the seeming imperfection of the animal itself, it was not to be supposed that they should be testicles; for then the animal should have partaken of the bull, which it certainly did not."

"This animal cannot be said to have been a mixture of all the parts of both sexes, for the clitoris had nothing similar to the penis in the male, and it was deficient in the female parts, by having nothing similar to the ovaria; neither had the uterus a cavity."

Works, vol. iv, p. 41.

BIOGRAPHICAL NOTES

COWPER, William (1666-1709)
Anatomist and Surgeon. F.R.S. 1698, the year in which his *Anatomy of Humane Bodies* was published. For this great work he "borrowed" the illustrations from Godfrey Bidloo's *Anatomy Atlas* of 1685 without, however, acknowledging his indebtedness. He described the glandulae bulbo-urethrales in 1697 in the *Phil. Trans.* (**21**. 364-369). Page 194.

FALLOPIUS, Gabriele (1523-1563).
Anatomist and Surgeon. Studied medicine in Ferrara and in Pisa; in 1551 was appointed Professor in Padua. His *Observationes anatomicae* was published in Venice in 1561. Pages 210-213

MORGAGNI, Giovanni Battista (1682-1771).
Professor of Anatomy in Padua for 59 years; regarded as the founder of morbid anatomy. Pages 204, 210, 212.

MÜLLER, Johannes Peter (1801-1858).
Professor of Anatomy in Bonn and later in Berlin. He described the duct in his work *Bildungsgeschichte der Genitalien aus anatomischen Untersuchungen an Embryonen des Menschen und der Thiere*, published in Dusseldorf in 1830 (see p. 60). Pages 204, 212.

WOLFF, Kaspar Friedrich (1733-1794).
Professor of Anatomy and Physiology in St Petersburg. His *Theoria generationis* was published in Halle in 1759. Page 204.

WORM, Ole (1588-1654).
Professor of Greek, Philosophy and Anatomy in Copenhagen. The bones were so named by Thomas Bartholin Primus, his nephew by marriage. Page 3.

INDEX

All the specimens are marked with letter 'P' before the number.

Printed by R. & R. Clark Ltd., Edinburgh